ENERGY SCIENCE, ENGINEERING AND TECHNOLOGY

FURTHER EXPLORATIONS IN THE ENERGY-WATER NEXUS

Energy Science, Engineering and Technology

Additional books in this series can be found on Nova's website under the Series tab.

Additional E-books in this series can be found on Nova's website under the E-book tab.

Environmental Science, Engineering and Technology

Additional books in this series can be found on Nova's website under the Series tab.

Additional E-books in this series can be found on Nova's website under the E-book tab.

ENERGY SCIENCE, ENGINEERING AND TECHNOLOGY

FURTHER EXPLORATIONS IN THE ENERGY-WATER NEXUS

DIANE A. HUGHES
AND
BRIAN F. SMITH
EDITORS

Nova Science Publishers, Inc.
New York

Library of Congress Cataloging-in-Publication Data

ISBN: 978-1-62081-318-8

Published by Nova Science Publishers, Inc. †New York

CONTENTS

Preface		**vii**
Chapter 1	Energy-Water Nexus: Information on the Quantity, Quality, and Management of Water Produced during Oil and Gas Production *United States Government Accountability Office*	**1**
Chapter 2	Produced Water Volumes and Management Practices in the United States *C. E. Clark and J. A. Veil*	**49**
Chapter 3	Energy-Water Nexus: Amount of Energy Needed to Supply, Use, and Treat Water Is Location-Specific and Can Be Reduced by Certain Technologies and Approaches *United States Government Accountability Office*	**107**
Chapter 4	A Review of Operational Water Consumption and Withdrawal Factors for Electricity Generating Technologies *Jordan Macknick, Robin Newmark, Garvin Heath and KC Hallett*	**139**
Index		**161**

PREFACE

The exploration for and production of oil and gas to meet our nation's energy needs also results in the production of large quantities of water as a byproduct. This water, which is produced from wells during exploration and production, is known as "produced water." Because produced water may contain a variety of contaminants, such as salts and minerals, it is often considered to be a waste stream that oil and gas producers must appropriately manage and treat before this water can be disposed of. If it is not appropriately managed or treated, the contaminants present in produced water discharged from oil and gas operations may threaten human health and the environment. This book explores the inextricable link between energy production and water with a focus on what is known about the volume and quality of produced water from oil and gas production; what practices are generally used to manage and treat produced water; and how the management of produced water is regulated at the federal level and in selected states.

Chapter 1 - Water is a significant byproduct associated with oil and gas exploration and production. This water, known as "produced water," may contain a variety of contaminants. If produced water is not appropriately managed or treated, these contaminants may present a human health and environmental risk.

GAO was asked to describe (1) what is known about the volume and quality of produced water from oil and gas production; (2) what practices are generally used to manage and treat produced water, and what factors are considered in the selection of each; (3) how produced water management is regulated at the federal level and in selected states; and (4) what federal research and development efforts have been undertaken during the last 10 years related to produced water. To address these objectives, GAO reviewed studies and other documents on produced water and interviewed federal and state regulatory officials, federal scientists, officials from oil and gas companies and water treatment companies, and other experts. GAO focused its review on the nine states that generate nearly 90 percent of the produced water, and conducted site visits in three states.

Chapter 2 - Produced water is water from underground formations that is brought to the surface during oil or gas production. Produced water is the largest volume by-product or waste stream associated with oil and gas exploration and production. The cost of managing such a large volume of water is a key consideration to oil and gas producers. Previous national produced water volume estimates are in the range of 15 to 20 billion barrels (bbl; 1 bbl = 42 U.S. gallons) generated each year in the United States. This is equivalent to a volume of 1.7 to 2.3 billion gallons per day. In comparison, the Washington D.C. government and surrounding jurisdictions provide about 300 million gallons per day of drinking water to local residents, businesses, and other users. This represents only about 13% of the daily produced water volume. As another example, consider a backyard swimming pool that is 20 ft wide by 50 ft long and 5 ft deep. The volume of water needed to fill such a pool is about 37,000 gal or about 900 bbl.

Produced water volume generation and management in the United States are not well characterized at a national level. The U.S. Department of Energy asked Argonne National Laboratory to compile data on produced water associated with oil and gas production to better understand the production volumes and management of this water.

This report provides a current estimate for the volume of produced water generated from oil and gas production in the United States. The volume estimate represents a compilation of data obtained from numerous state oil and gas agencies and several federal agencies for 2007, where possible. The total volume of produced water estimated for 2007 is about 21 billion bbl. This equals an average of 57.4 million bbl/day. Produced water is generated from most of the nearly 1 million actively producing oil and gas wells in the United States.

Chapter 3 - Providing drinking water and wastewater services are two key functions needed to support an urban lifestyle. To provide these services, energy is needed to extract, use, and treat water and wastewater. As the demand for water increases, the energy demands associated with providing water services are similarly expected to grow.

GAO was asked to describe what is known about (1) the energy needed for the urban water lifecycle and (2) technologies and approaches that could lessen the energy needed for the lifecycle and barriers that exist to their adoption. To address these issues, GAO reviewed scientific studies, government-sponsored research, and other reports and interviewed specialists from a variety of organizations, including drinking water and wastewater utilities; federal, state, and local government offices responsible for water or energy; and relevant nonprofit groups, about the energy needed to move, use, and treat water. GAO also selected three cities—Memphis, Tennessee; San Diego, California; and Washington, D.C.—as illustrative case studies to help understand the energy demands of the lifecycle in different areas of the country.

GAO is not making any recommendations in this report. A draft was provided to the Departments of Defense, Energy (DOE), and the Interior, and the Environmental Protection Agency (EPA). DOE and EPA provided technical comments, which we incorporated as appropriate.

Chapter 4 - This report provides estimates of operational water withdrawal and water consumption factors for electricity generating technologies in the United States. Estimates of water factors were collected from published primary literature and were not modified except for unit conversions. The presented water factors may be useful in modeling and policy analyses where reliable power plant level data are not available. Major findings of the report include:

- The power sector withdraws more water than any other sector in the United States and is heavily dependent on available water resources. Changes in water resources may impact the reliability of power generation.
- Water withdrawal and consumption factors vary greatly across and within fuel technologies. Water factors show greater agreement when organized according to cooling technologies as opposed to fuel technologies. Once-through cooling technologies withdraw 10 to 100 times more water per unit of electric generation than recirculating cooling technologies; recirculating cooling technologies consume at least twice as much water as once-through cooling technologies.
- A transition to a less carbon-intensive electricity sector could result in either an increase or decrease in water use, depending on the choice of technologies and cooling systems employed. Concentrating solar power (CSP) technologies and coal facilities with carbon capture and sequestration (CCS) capabilities have the highest water consumption values when using a recirculating cooling system. Non-thermal

renewables, such as photovoltaics (PV) and wind, have the lowest water consumption factors.

- Federal datasets on water use in power plants have numerous gaps and methodological inconsistencies. Federal agencies are currently coordinating to improve these data. Water use factors discussed here are good proxies for use in modeling and policy analyses, at least until power plant level data improve.
- Impacts of the power sector on freshwater availability can be reduced by utilizing dry cooling or by using non-freshwater sources for cooling. However, these alternatives are limited by locally available resources and may have cost and performance penalties.

Improved power plant data and further studies into the water requirements of energy technologies in different climatic regions would facilitate greater resolution in analyses of water impacts of future energy and economic scenarios. This report provides the foundation for conducting water use impact assessments of the power sector while also identifying gaps in data that could guide future research.

In: Further Explorations in the Energy-Water Nexus ISBN 978-1-62081-318-8
Editors: D. A. Hughes and B. F. Smith

Chapter 1

ENERGY-WATER NEXUS: INFORMATION ON THE QUANTITY, QUALITY, AND MANAGEMENT OF WATER PRODUCED DURING OIL AND GAS PRODUCTION*

United States Government Accountability Office

WHY GAO DID THIS STUDY

Water is a significant byproduct associated with oil and gas exploration and production. This water, known as "produced water," may contain a variety of contaminants. If produced water is not appropriately managed or treated, these contaminants may present a human health and environmental risk.

GAO was asked to describe (1) what is known about the volume and quality of produced water from oil and gas production; (2) what practices are generally used to manage and treat produced water, and what factors are considered in the selection of each; (3) how produced water management is regulated at the federal level and in selected states; and (4) what federal research and development efforts have been undertaken during the last 10 years related to produced water. To address these objectives, GAO reviewed studies and other documents on produced water and interviewed federal and

* This is an edited, reformatted and augmented version of the Highlights of GAO-12-156, a report to the Ranking Member, Committee on Science, Space, and Technology, House of Representatives, dated January 2012.

state regulatory officials, federal scientists, officials from oil and gas companies and water treatment companies, and other experts. GAO focused its review on the nine states that generate nearly 90 percent of the produced water, and conducted site visits in three states.

What GAO Recommends

GAO is not making any recommendations. A draft was provided to the Departments of Energy and the Interior, and EPA for review. None of these agencies provided written comments. EPA and Interior provided technical comments, which we incorporated as appropriate.

What GAO Found

A significant amount of water is produced daily as a byproduct from drilling of oil and gas. A 2009 Argonne National Laboratory study estimated that 56 million barrels of water are produced onshore every day, but this study may underestimate the current total volume because it is based on limited, and in some cases, incomplete data generated by the states. In general, the volume of produced water generated by a given well varies widely according to three key factors: the hydrocarbon being produced, the geographic location of the well, and the method of production used. For example, some gas wells typically generate large volumes of water early in production, whereas oil wells typically generate less. Generally, the quality of produced water from oil and gas production is poor, and it cannot be readily used for another purpose without prior treatment. The specific quality of water produced by a given well, however, can vary widely according to the same three factors that impact volume—hydrocarbon, geography, and production method.

Oil and gas producers can choose from a number of practices to manage and treat produced water, but underground injection is the predominant practice because it requires little or no treatment and is often the least costly option. According to federal estimates, more than 90 percent of produced water is managed by injecting it into wells that are designated to receive produced water. A limited amount of produced water is disposed of or reused by producers in other ways, including discharging it to surface water, storing it in surface impoundments or ponds so that it can evaporate, irrigating crops, and reusing it for hydraulic fracturing. Managing produced water in these ways can require more advanced treatment methods, such as distillation. How produced water is ultimately managed and treated is primarily an economic decision, made within the bounds of federal and state regulations.

The management of produced water through underground injection is subject to the Safe Drinking Water Act's Underground Injection Control program, which is designed to prevent contamination of aquifers that supply public water systems by ensuring the safe operation of injection wells. Under this program, the Environmental Protection Agency (EPA) or the states require producers to obtain permits for their injection wells by, among other things, meeting technical standards for constructing, operating, and testing and monitoring the wells. EPA also regulates the management of produced water through surface discharges under the Clean Water Act. Other management practices, such as disposal of the water into surface impoundments, irrigation, and the reuse of the water for hydraulic fracturing, are regulated by state authorities.

Several federal agencies, including EPA; the Department of Interior's Bureau of Reclamation and U.S. Geological Survey; and a number of Department of Energy national laboratories, have undertaken research and development efforts related to produced water. These efforts have included sponsoring and issuing studies that describe the volume and quality of produced water, options for managing produced water and associated regulatory issues, as well as options for improving existing technologies for treating produced water and developing new technologies, such as more cost-effective filters.

ABBREVIATIONS

BLM	Bureau of Land Management
DOE	Department of Energy
EIA	Energy Information Administration
EPA	Environmental Protection Agency
NETL	National Energy Technology Laboratory
USGS	U.S. Geological Survey

January 9, 2012

The Honorable Eddie Bernice Johnson
Ranking Member
Committee on Science, Space, and Technology
House of Representatives

Dear Ms. Johnson:

The exploration for and production of oil and gas to meet our nation's energy needs also results in the production of large quantities of water as a byproduct.[1] This water, which is produced from wells during exploration and production, is known as "produced water." Because produced water may contain a variety of contaminants, such as salts and minerals, it is often considered to be a waste stream that oil and gas producers must appropriately manage and treat before this water can be disposed of. If it is not appropriately managed or treated, the contaminants present in produced water discharged from oil and gas operations may threaten human health and the environment.

Oil and gas—known as hydrocarbons—are found in a variety of geologic formations. Oil can be found in deep, porous rock or reservoirs that can flow under natural pressure to the surface after drilling, as well as other geologic formations, including shale and oil sands, in which other processes must be used to extract the oil, such as injecting liquid into the formation or applying heat or steam. Similarly, gas can be found in porous rock or reservoirs, in coal seams (known as coalbed methane), and in tighter geologic formations, including tight sands and shale formations. Extracting oil and gas from any of these reserves can result in produced water as a byproduct because water can exist naturally along with oil and gas in geologic formations (known as formation water) or it can be added to the well to stimulate oil and gas production (known as injected water). Formation water and injected water can return to the surface as produced water along with the oil or gas that is being extracted from the well. In other circumstances, such as when oil and gas are extracted from certain shale formations, water, sand, and chemicals are injected at high pressure to create fractures in the formation—a process known as hydraulic fracturing—thereby allowing the oil or gas to flow easier and be brought to the surface.[2] Some of this mix of water, sand, and chemicals returns to the surface when production starts (this type of produced water is known as flowback water). Over time, the quantity of produced flowback water generally diminishes, but smaller amounts of water will continue to be produced from the well.

Because of the inextricable link between energy production and water, you asked us to undertake a series of studies reviewing the energy-water nexus.[3] This is the fifth and final study in this series and provides information on (1) what is known about the volume and quality of produced water from oil and gas production; (2) what practices are generally used to manage and treat produced water, and what factors are considered in the selection of each; (3) how the management of produced water is regulated at the federal level and in selected states; and (4) what federal research and development efforts have been undertaken during the last 10 years related to produced water.

While oil and gas production comes from onshore and offshore operations, the focus of this review is onshore because produced water from onshore sources has the potential to affect surface and groundwater quality, and a greater variety of practices are employed to manage produced water. To address these objectives, we conducted a literature review of studies and other documents on produced water quality and volume, management, and regulations issued by federal agencies and laboratories, state agencies, the oil and gas industry, and academic institutions. These documents included peer-reviewed scientific and industry periodicals, government-sponsored research, and reports from nongovernmental research organizations. We believe we have included the key studies and have qualified our findings, where appropriate. However, it is possible that we may not have identified all of the studies with findings relevant to our objectives.

In addition, we interviewed stakeholders such as federal and state regulatory officials; federal scientists from the Environmental Protection Agency's (EPA) Office of Research and Development and the Department of Energy's (DOE) Argonne National Laboratory, Los Alamos National Laboratory, National Energy Technology Laboratory (NETL), Oak Ridge National Laboratory, and Sandia National Laboratories; officials from oil and gas exploration and production companies; officials from water treatment companies; and other experts with experience related to produced water. The federal and state regulatory officials included those with responsibility over oil and gas regulation, as well as clean water and drinking water regulation. We focused our review of management techniques and produced water regulation on nine states—California, Colorado, Kansas, Louisiana, New Mexico, Oklahoma, Pennsylvania, Texas, and Wyoming. These states account for nearly 90 percent of produced water generated.

We supplemented our literature review and stakeholder discussions with site visits to selected locations in Pennsylvania, Texas, and Wyoming, where we met with oil and gas producers and officials from produced water treatment facilities and discussed issues related to produced water management and treatment and the factors that influence these decisions. We selected these states because of the current and potential volumes of produced water generated, the range of hydrocarbons produced, and the different management and treatment practices employed. We also visited hydraulic fracturing drilling operations, underground injection control well sites, and a number of different treatment facilities employing a variety of technologies. To determine what federal research and development efforts have been undertaken during the last 10 years related to produced water, we analyzed information supplied by and conducted interviews with federal officials from DOE and select national laboratories, EPA, and the Department of the Interior's U.S. Geological Survey (USGS), Bureau of Land Management (BLM), and Bureau of Reclamation.

We conducted this performance audit from October 2010 to January 2012, in accordance with generally accepted government auditing standards. Those standards require that we plan and perform the audit to obtain sufficient, appropriate evidence to provide a reasonable basis for our findings and conclusions based on our audit objectives. We believe that the evidence obtained provides a reasonable basis for our findings and conclusions based on our audit objectives.

BACKGROUND

The domestic production of oil and gas is essential to the nation's energy portfolio. According to DOE's Energy Information Administration (EIA), the United States consumed approximately 6.7 billion barrels of oil and 24 trillion cubic feet of gas in 2010. Together, oil and gas production supply over 60 percent of the nation's total energy demand, and demand is expected to grow in the future. While domestic drilling for oil and gas can present risks to the environment, it also results in the creation of jobs and economic growth, as well as payments to the government in the form of royalties.[4]

The U.S. oil and gas reserves located onshore are extensive, with proven onshore oil reserves of approximately 18 billion barrels and proven onshore gas reserves of approximately 271 trillion cubic feet as of December 2009, according to the EIA.[5] These reserves are located across the country, primarily in 31 key oil- and gas-producing states, and include a range of different types of hydrocarbons. These hydrocarbons include both conventional and unconventional oil and gas sources. Although there is no clear and consistently agreed upon distinction between conventional and unconventional oil and gas, conventional sources of oil and gas are generally produced using traditional methods of drilling and pumping, whereas unconventional oil and gas sources generally require more complex and expensive technologies for production. For example, in some instances heavy oils produced using steam injection are considered unconventional, while in other situations they are considered conventional. For this reason, in this report we generally do not distinguish between conventional and unconventional oil and gas except in instances when supporting documentation or information we use from other sources makes this distinction necessary; in such cases, we provide the relevant definition in a footnote.[6]

Oil and gas is found within underground layers of rock referred to as formations. The geologic characteristics of the formations in which the various hydrocarbons are found vary widely, along with the characteristics of the hydrocarbons themselves. For example, shale oil and gas formations are generally tighter and much less permeable than other formations, causing the oil and gas to be much less free flowing. Coalbed methane formations, located at shallow depths of 1,000 to 2,000 feet, are more permeable formations through which gas can flow more freely than through shale formations. In addition, heavy oil, due to its higher viscosity, has much less ability to flow freely through a formation compared to lighter oil.[7] Figure 1 below provides examples of differing geologies for various gas hydrocarbons.

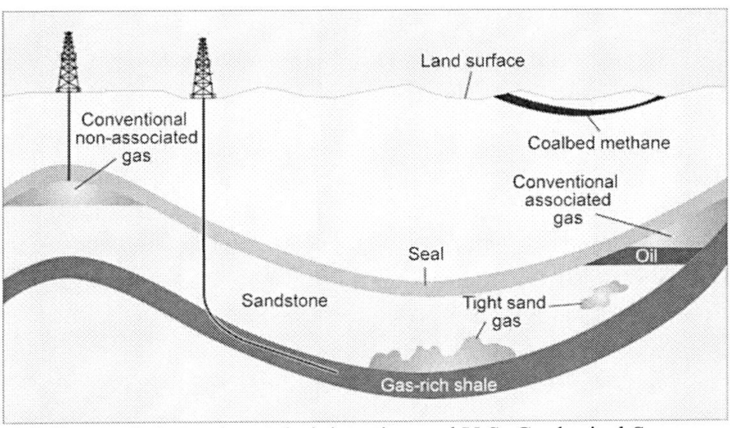

Source: U.S. Energy Information Administration and U.S. Geological Survey.
Note: "Associated gas" is gas that accumulates in conjunction with oil in a formation. "Non-associated gas" is gas that accumulates separately from oil in a formation.

Figure 1. Geology of Gas Resources.

As a result of these geologic differences, the methods used to produce hydrocarbons vary widely. Some oil and gas can be produced by drilling a well and relying on the natural pressure in the formation to push the oil or gas to the surface. Heavy oil, on the other hand, may require the injection of an additive such as steam into the formation to stimulate the flow of oil—a process known as enhanced recovery. Similarly, some hydrocarbons are produced through the use of hydraulic fracturing. Hydraulic fracturing involves the injection of liquid under pressure to fracture the rock formation and to prop open the fractures to allow hydrocarbons to flow more freely from the formation into the well for collection. The liquids used in this process consist primarily of water, but also include chemicals, as well as sand or other propping agents for holding open the fractures (proppant). Hydraulic fracturing is commonly used to facilitate the production of many hydrocarbons, including oil, shale gas, tight gas, and coalbed methane. Recent improvements in hydraulic fracturing, combined with horizontal drilling technologies, have prompted a boom in shale oil and gas production.

The process of producing oil and gas is complicated and yields several byproducts that must be managed as part of the oil and gas operation's waste stream. Key among these byproducts is the produced water that comes to the surface along with the oil or gas during production. Produced water may include water that occurs naturally in the formation, water or other liquids that were injected into the formation to enhance recovery during the drilling or production process, and flowback water, which is the water, proppants, and chemicals used for hydraulic fracturing (fracturing fluids).

EPA regulates water primarily through two federal laws: the Safe Drinking Water Act and the Clean Water Act. The Safe Drinking Water Act was originally passed by Congress in 1974 to protect public health by ensuring a safe drinking water supply.[8] Under the act, EPA is authorized to set standards for both naturally-occurring and man-made contaminants that may be found in drinking water. The Safe Drinking Water Act also regulates the placement of wastewater and other fluids underground through the Underground Injection Control program. This program provides safeguards to ensure that wastewater injected underground does not endanger drinking water supplies. There are six classes or categories of wells regulated through the Underground Injection Control program. For example, class II wells are for the management of fluids associated with oil and gas production, and they include wells used to dispose of wastewater and those used to enhance oil and gas production.[9] EPA may grant, or approve by rule, primary enforcement authority for the Underground Injection Control program to a state, which means that the state assumes responsibility for executing the program, including permitting, monitoring, and enforcement for operations within the state. To be approved for this authority, state programs must be at least as stringent as the federal program and show that their regulations contain effective minimum requirements. To obtain this authority over class II wells only, states with existing oil and gas programs may make an optional demonstration that their program is effective in protecting underground sources of drinking water.

The Clean Water Act was enacted by Congress in 1972 to protect surface waters by regulating discharges of pollutants into those waters.[10] Pursuant to the water-quality-based pollution control program mandated by the act, states establish and EPA approves water quality standards for contaminants in surface waters. The Clean Water Act also regulates the discharge of wastewaters, including produced water, through the National Pollutant Discharge Elimination System permit program, which requires all facilities that discharge pollutants from any point source into surface waters to be permitted.[11] EPA may delegate primary enforcement authority to a state for this program if a state demonstrates that its program requirements are as stringent as those set by EPA. Once EPA delegates this authority, the state is responsible for permitting, monitoring, and enforcing these permits.

Oil and gas producers who would like to drill on federal lands managed by BLM must also obtain permits from BLM. The Federal Land Policy and Management Act of 1976 requires BLM to develop resource management plans, which, among other things, identify parcels of federal land that will be available for oil and gas development and leasing. Producers who have obtained a lease from BLM for oil and gas production on public lands must submit an application to BLM for a permit to drill before beginning to prepare the land or drilling any new oil or gas wells. The complete permit application package must include, among other things, a reclamation plan that details the steps operators propose to take to reclaim the site, including redistribution of topsoil, configuring the reshaped topography, and seeding or other steps to re-establish vegetation. In some circumstances, approval from state officials may also be required before operators can begin drilling and production.

In addition to obtaining permits for underground injection, discharge, and drilling, oil and gas producers may also be required to obtain permits from state or regional authorities to withdraw water from a river basin or underground. For example, producers in the Susquehanna River Basin must obtain a permit from the Basin Commission to withdraw surface water or groundwater. This water may be blended with recycled or produced water and used for hydraulic fracturing. Alternately, producers in Colorado and

Wyoming must obtain groundwater permits from their State Engineer's Office to drill for coalbed methane. State officials consider the drilling to be a beneficial use of water since producers must withdraw water from underground to obtain the gas and this classification means a permit is required.

Oil and Gas Wells Generate a Significant Amount of Produced Water, but the Volume and Quality of the Water Produced at a Given Well Varies

A significant amount of water is produced daily as a byproduct from onshore drilling of oil and gas, but the volume produced by a given well will vary depending on the type of hydrocarbon being produced, the geographic location of the well, and the method of production used. Overall, most produced water is of poor quality and cannot be used for other purposes without prior treatment; however, produced water quality can also vary greatly depending on the hydrocarbon, geography, and production method.

Millions of Barrels of Produced Water Are Generated Daily, but the Volume Produced at a Given Well Is Dependent on Several Factors

An estimated 56 million barrels of produced water are generated every day as a byproduct of onshore oil and gas production in the United States. This estimate is based on an Argonne National Laboratory study of produced water volumes generated during 2007—the most recent year for which such data were collected—and was derived from data collected from state agencies in 31 oil- and gas-producing states.[12] The study is considered by agency officials, researchers, and other experts with whom we spoke to be the most comprehensive and accurate assessment of produced water volumes to date. However, because the Argonne study is based on limited and, in some cases, incomplete data, it likely underestimates the current total volume of produced water being generated by oil and gas operations today. Specifically, we noted the following limitations in the Argonne estimate:

- The reporting time frame for the study largely occurred prior to the recent, dramatic increase in shale gas production in the United States, which had an average annual growth rate of 48 percent from 2006 to 2010, according to the EIA. In 2010, EIA had estimated that shale gas accounted for approximately 23 percent of all gas production in the United States.
- The study is based on state data that were collected and maintained using a variety of methods, making them difficult to compare and aggregate at a national level. For example, in some states, producers are required to report produced water volumes as measured by flow meters, whereas in other states, producers are required to estimate produced water volumes using a method of their choosing, which may limit the precision of the data. Furthermore, in other states, producers are only required to report produced water volumes they dispose of in injection wells, which may not reflect the total produced water generated. Lastly, some states do not collect data on produced water volumes at all. In such cases, Argonne generated estimates of produced water volumes based on available information on oil and gas production in the state and made certain assumptions about water volumes based on produced water data from neighboring states.

Although the Argonne study clearly demonstrates that a large amount of produced water is generated daily, the volume generated by a specific oil and gas well can vary significantly according to three key factors: the type of hydrocarbon being produced, the geographic location of the well, and the method of production used. First, according to literature we reviewed and stakeholders we spoke with, the type of hydrocarbon influences not only how much water a well generates, but also when the water is produced over the life of the well. This is because the geological formations for different hydrocarbons have different attributes, thus influencing the amount of water that is produced from a particular well. For example, coalbed methane wells produce large volumes of water in the early stages of production, because coal beds are essentially aquifers that contain coal rock and gas bound together from the pressure of the water present in the aquifer.[13] By pumping out water, the resulting drop in pressure allows the gas to detach from the coal and flow to the surface. In contrast, one producer noted that their conventional gas wells produce much less water than their coalbed methane wells because the formations from which conventional gas is drawn contain much less water.[14] Oil wells, on the other hand, typically generate less water during the first years of production, when formation pressure is high enough to allow the oil to flow freely to the surface. As these oil wells age, however, water volumes increase, as oil taken out is displaced by water flowing in from the surrounding formation. One producer that we spoke with noted that their older oil fields produce more than five times the volume of water produced by their younger oil fields.

Second, according to the literature we reviewed and stakeholders we spoke with, the geographic location of a well also influences the volume of produced water it generates, due to differences in geology. For example, stakeholders noted that the Barnett Shale formation in Texas is generally known to be a "wetter" formation than the Marcellus Shale formation in the Northeast, with shale gas wells in the Barnett typically producing three to four times more water than shale gas wells in the Marcellus. Similarly, USGS reported that coalbed methane wells in the Powder River Basin in Wyoming and Montana produce, on average, 16 times more water than coalbed methane wells in the San Juan Basin in Colorado and New Mexico. In addition, produced water volumes can vary among wells in close proximity with one another. For example, at one site we visited in Wyoming, some gas wells were producing two to three times more water than other gas wells in the same field, for reasons that were, in general, not clear to the operator of those wells.

Lastly, the method of production used to extract oil and gas also influences the volume of water generated, according to the literature we reviewed and stakeholders we spoke with. Specifically, stakeholders reported that methods of production that rely on the injection of water and other fluids into the formation in order to stimulate oil and gas production can generate more produced water than in cases in which oil and gas comes to the surface under existing pressure. For example, one stakeholder reported that the use of enhanced oil recovery methods such as steam injection can generate eight to nine barrels of water for every barrel of oil produced. At one enhanced oil recovery field we visited, produced water comprised more than 95 percent of the total liquids produced, with oil comprising the remainder. Similarly, the use of hydraulic fracturing to produce oil or gas can result in larger volumes of produced water than production in more porous formations, although the larger volumes associated with hydraulic fracturing are limited to the initial flowback of water and fracturing fluids. For example, with shale gas production, stakeholders reported that flowback volumes can range from approximately 10,000 to 60,000 barrels per well for each hydraulic fracture.[15]

However, once the initial flowback ceases, the volume of water produced by shale gas production may be relatively small, sometimes decreasing to just a few barrels per day.

Produced Water Is Generally of Poor Quality, with the Levels of Contaminants Varying Widely

The quality of produced water from oil and gas production is generally poor, and in most situations, it cannot be readily used for other purposes without prior treatment. According to the literature we reviewed and stakeholders we spoke with, produced water may contain a wide range of contaminants in varying amounts. Most of the contaminants occur naturally in the produced water, but some are added through the process of drilling, hydraulic fracturing, and pumping oil and gas. The range of contaminants found in produced water can include, but is not limited to

- salts, which include chlorides, bromides, and sulfides of calcium, magnesium, and sodium;
- metals, which include barium, manganese, iron, and strontium, among others;
- oil, grease, and dissolved organics, which include benzene and toluene, among others;
- naturally occurring radioactive materials; and
- production chemicals, which may include friction reducers to help with water flow, biocides to prevent growth of microorganisms, and additives to prevent corrosion, among others.

Exposure to these contaminants at high levels may pose risks to human health and the environment. For example, according to EPA, a potential human health risk from exposure to high levels of barium is increased blood pressure, and potential human health risks from exposure to high levels of benzene are anemia and increased risk of cancer. From an environmental standpoint, research indicated that elevated levels of salts can inhibit crop growth by hindering a plant's ability to absorb water from the soil. Additionally, exposure to elevated levels of metals and production chemicals, such as biocides, can contribute to increased mortality among livestock and wildlife.

The specific quality of water generated by a given well, however, can vary widely according to the same three factors that impact the volume of water produced from the well: the hydrocarbon being produced, the geographic location of the well, and method of production used. First, according to stakeholders we spoke with, the type of hydrocarbon is a key driver of produced water quality, due to differences in geology across the formations in which the hydrocarbons are found. Specifically, the depth at which the hydrocarbons are found influences the salt and mineral content of produced water, and, in general, the deeper the formation is, the higher the salt and mineral content will be. For example, produced water from shale gas wells drilled at depths generally ranging from 5,000 to 8,000 feet have salt and mineral levels 20 times higher than produced water from coalbed methane wells drilled at depths of 1,000 to 2,000 feet. Additionally, the amount of oil or gas that is mixed in with the produced water brought to the surface can also vary. For example, produced water typically blends more easily with oil than with gas. As a result, produced water from oil wells generally contains levels of oil, grease, and other organic compounds that are four to five times higher than water from gas wells.

Second, the quality of produced water also varies depending on the well's geographic location, also because of differences in geology. For example, producers we spoke with said that produced water from wells in the Marcellus Shale formation in the Northeast has higher levels of radionuclides than water from shale gas wells in the Barnett Shale formation in Texas. Similarly, according to research, produced water from coalbed methane wells in the Raton Basin in Colorado and New Mexico has a salt content, on average, roughly two and a half times higher than produced water from the Powder River Basin in Wyoming and Montana. In addition, produced water quality can vary within a given region, according to producers we spoke with. For example, some coalbed methane wells in the Powder River Basin in Wyoming contain barium levels five to six times higher than the barium levels found in wells less than 50 miles away. Additionally, produced water from wells in one oil field in California contains levels of boron four to five times higher than produced water from oil wells in neighboring fields.

Lastly, the method of production can affect the quality of the water produced. These differences are largely attributable to the chemicals and other substances added during drilling or production processes, according to stakeholders we spoke with. Specifically, methods of production that rely on hydraulic fracturing or enhanced recovery methods can result in poorer quality produced water than other methods. For example, according to stakeholders, the range of chemicals, sand, and water that are added to facilitate the hydraulic fracturing process can lower the overall quality of the produced water from these kinds of operations. Similarly, the use of chemicals during enhanced recovery can also affect the quality of water produced. Stakeholders noted that enhanced recovery involves the addition of production chemicals such as biocides, corrosion inhibitors, and friction reducers, along with steam or carbon dioxide. For example, one stakeholder estimated that wells produced using these enhanced recovery methods can yield produced water with levels of some production chemicals three to four times higher than produced water from wells that do not use enhanced recovery techniques.

A Number of Practices Are Available to Manage and Treat Produced Water, with Cost Being the Primary Determining Factor

Oil and gas producers have a number of options on how to manage produced water, but underground injection is the predominant practice. In addition to underground injection, a limited amount of produced water is managed by discharging it to surface water, storing it in surface impoundments, and reusing it for irrigation or hydraulic fracturing. With regard to treatment options, most produced water is minimally treated, although more advanced treatment methods are available if the end use of the water requires a higher level of treatment. Ultimately, cost is the primary driver in producers' decisions about how to manage and treat produced water generated by oil and gas producers.

Produced Water Can be Managed in a Number of Ways, although Underground Injection Is the Most Common Practice

Over 90 percent of the water produced during oil and gas operations is managed through underground injection practices; the remaining water is generally discharged to surface water, stored in surface impoundments, reused for irrigation, or reused for hydraulic fracturing. In its 2009 report,[16] Argonne National Laboratory estimated, and EPA officials that we spoke with concurred, that most produced water is managed by injecting it underground into wells that are designated to receive this water.[17] These wells, known as injection wells, must be constructed to protect underground sources of drinking water, and they are tested and monitored periodically to ensure no drinking water is being contaminated by well operations (see figure 2). Injection wells can be used for enhanced recovery or permanent disposal of the water. When producers reuse produced water for enhanced recovery, they inject it into wells in the same producing formation to recover additional oil and, in limited applications, gas, thus prolonging the life of the production well. Some of this water will come back up as produced water in subsequent well operations. When producers inject produced water for permanent disposal into an underground formation, they inject it into wells in the same formation or a formation that is similar to the one the produced water was extracted from.

According to EPA records, in 2010 there were 150,855 injection wells authorized for the injection of fluids brought to the surface during oil and gas production, including produced water, although EPA officials told us that not all are currently operating.[18] About four-fifths of the wells—124,837—are located in the nine states we reviewed (see table 1).

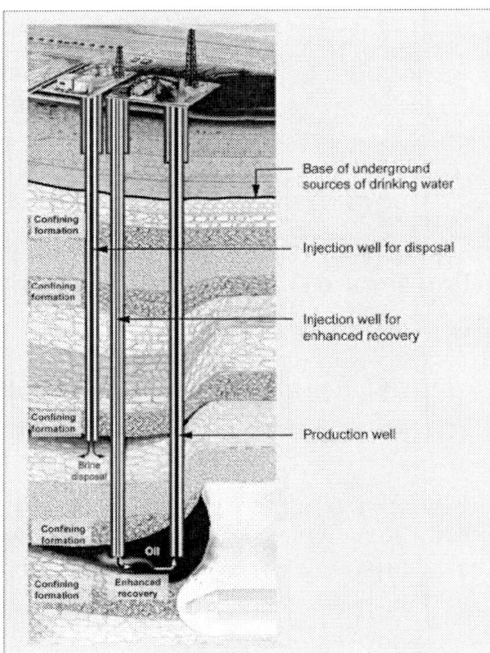

Source: EPA.

Figure 2. Injection Wells for Produced Water.

Table 1. Number of Injection Wells in Selected States

State	Number of injection wells
Texas	52,016
California	29,505
Kansas	16,658
Oklahoma	10,629
Wyoming	4,978
New Mexico	4,585
Louisiana	3,731
Pennsylvania	1,861
Colorado	874
Total in selected states	**124,837**

Source: GAO analysis of EPA data.

According to state regulators we interviewed in the nine states, underground injection is common in most, but not all, of their states. Specifically, regulators in five states told us that all or almost all of the produced water is managed through underground injection, and in three other states, most of the produced water is managed this way. In the ninth state—Pennsylvania—many producers use underground injection for enhanced recovery, but the practice is not widely used for disposal, according to EPA officials.[19] Produced water that is not injected into underground injection wells is disposed of or reused by producers in other ways, including the following.[20]

Discharge to surface water. According to Argonne National Laboratory's report, less than 1 percent of produced water generated from onshore oil and gas operations in 2007 was managed by discharging it to surface water.[21] Surface discharges of produced water directly from oil and gas production sites are prohibited in much of the United States, but produced water may be discharged from an off-site treatment facility.[22] According to Argonne's report, discharges of produced water to surface water are primarily limited to the western United States and generally occur only when the salt content of the water is low. While a current national estimate of this practice is not available, EPA officials and regulators in seven of the nine states we reviewed told us that surface discharges of produced water are limited or are nonexistent in their states. For example, officials we spoke with from Colorado said that a very small portion of the produced water generated in that state is discharged to surface water and only 24 of the approximately 9,900 discharge permits they have issued are for oil and gas producers. Among the states we reviewed, Wyoming and Pennsylvania were the only two where producers commonly use surface discharges to manage produced water.[23] For example, an oil producer we spoke with in Wyoming told us his company discharges a small portion of the produced water from some of its fields directly to a nearby creek because the water quality is high enough to meet the state's discharge limits without prior treatment; however, the majority of the water generated from these fields has a much higher salt concentration and cannot be discharged into surface waters.

Surface impoundment. Surface impoundments are lined or unlined ponds used primarily to facilitate evaporation of produced water, and in the case of unlined ponds, allow it to infiltrate into the ground.[24] According to DOE's NETL, drier climates are favorable for evaporation and spray nozzles may be used to increase the rate of evaporation. The National Research Council reported that, in 2008, surface impoundments were used to manage about 64 percent of produced water generated by coalbed methane producers in the Wyoming portion of the Powder River Basin.[25] According to the report, the water was generally untreated, but in some cases water was treated to meet requirements for discharging to the impoundment. For example, a coalbed methane producer we visited in Wyoming treats the produced water at a treatment facility to first remove barium to meet state water quality standards, then disposes of most of it in a surface impoundment, where it evaporates or infiltrates to the subsurface. Officials we spoke with from California, Colorado, and New Mexico also said that surface impoundments are used to manage produced water in those states, but it is not a significant practice in any of them.

Irrigation. Some produced water from coalbed methane is reused for irrigation in certain parts of Wyoming and Colorado because the water is generally of high enough quality that it does not require extensive treatment in order to avoid damage to the crops or soil. According to the National Research Council,[26] about 13 percent of the produced water generated from coalbed methane producers in Wyoming's Powder River Basin was reused for managed irrigation or subsurface drip irrigation.[27] For example, a coalbed methane production operation we visited in Wyoming disposes of almost all of its produced water from the Powder River Basin using a managed irrigation system, following minimal treatment of the water. Because its wells produce more water than can be disposed of under its surface discharge permit, it is a fairly economical option, and is allowable under state regulations. While there are examples of irrigation with produced water occurring elsewhere, it is not a widely used management practice. According to NETL, a significant challenge to using produced water for irrigation is the salt content of the water, which can decrease crop yields and damage the soil. In addition, the National Research Council reported that while reusing coalbed methane produced water for beneficial purposes such as irrigation would seem to be a desirable and relatively easy objective, in reality it is potentially economically and environmentally burdensome, complex, and challenging. The suitability of water for irrigation depends on a number of factors including the type of crops grown, the soil type, irrigation methods, and the types and quantity of salts dissolved in the water. In addition, the reliability of the produced water supply over time, proximity to the irrigation site, and costs also present challenges.

Hydraulic fracturing. In recent years, some shale gas producers have begun reusing produced water for hydraulic fracturing of additional wells at their operations. The water is typically treated first, either on-site or off-site, and then mixed with freshwater if salt concentrations remain high. Although no national estimate of producers' use of this practice is available, a 2009 report on shale gas development reported that interest in this type of reuse for produced water was high.[28] However, the report also noted that certain water treatment challenges needed to be overcome to make this type of reuse more widespread. According to NETL, in order for reuse of produced water to become widespread, low-cost treatment technologies must be developed. In the last couple of years, reusing produced water for hydraulic fracturing has become more common among shale gas producers in Pennsylvania, according to state regulators and producers we spoke with in the state. The shift was motivated, in part, by a change in the state's surface discharge standards that ultimately made treatment and discharge a comparatively more expensive practice. For example, one shale gas production site we visited in Pennsylvania currently reuses all of its produced water for hydraulic fracturing, although it had used other practices in the past. Other shale gas producers in the state are also adopting this approach, according to agency officials and an academic expert we spoke with. In addition, regulators in five of the other states we reviewed told us that producers in these states are reusing produced water for hydraulic fracturing, although generally to a lesser extent than in Pennsylvania.

Most Produced Water Is Minimally Treated, but More Advanced Treatment Methods Are Available

Because most produced water is managed through underground injection wells, it is minimally treated; however, if produced water is going to be reused or disposed of in some other manner, then more advanced treatment methods are available, depending on the level of treatment required.

Treatment methods for produced water managed through underground injection. Produced water managed through underground injection generally does not need to be treated because injection wells are designed to confine the produced water to the receiving formation and prevent it from migrating to underground sources of drinking water. In some cases, however, to meet an injection well's operating requirements or prevent premature "plugging" of the formation, the water may be treated to control excessive solids, dissolved oil, corrosion, chemical reactions, or the growth of bacteria and other microbes, according to NETL. Such treatment is generally minimal and can include storing the water in a tank to allow solids to settle out and passing the water through a screen or filter to remove additional solids. Chemicals may also be added to prevent corrosion of the injection well equipment and filtration or biocides may be used to prevent bacteria, algae, or fungi present in the water from clogging equipment or encouraging corrosion.

Treatment methods for produced water reused for hydraulic fracturing. Producers who reuse produced water for hydraulic fracturing told us they treat the water to meet their own operating requirements. While producers we spoke with said that they had previously treated the water to a very high quality before reusing it for hydraulic fracturing, they are currently experimenting with lower levels of treatment. For example, one producer told us that 2 years ago his company treated the water so that it was nearly as clean as freshwater, but based on internal research, the company no longer removes salt from the produced water that it reuses for hydraulic fracturing. This lower level of treatment has reduced operating costs, and the producer is considering eliminating other treatment steps as long as doing so will not cause operational problems, such as equipment corrosion.

Treatment methods for produced water discharged to surface water bodies or reused for irrigation. If produced water is going to be discharged to surface water or reused for irrigation, then treatment is often necessary to reduce hardness, salts, and other contaminants, in addition to settling and filtration methods to remove solids. Solids and hardness removal are sometimes referred to as "pretreatment" steps. Hard water contains dissolved constituents, mainly calcium and magnesium ions, which can cause scaling of pipes and equipment. Hardness is typically removed prior to removing salts by adjusting the pH of the water and adding chemicals that cause dissolved calcium and magnesium to form small solids, or precipitates, which then settle out or are filtered out of the water with the aid of additional processes.[29] Alternatively, when produced water is going to be reused for irrigation, calcium or magnesium may be added to the water to address sodium levels. Treatment technologies, including distillation, reverse osmosis, and ion exchange, are then used to remove salt and other contaminants from produced water.[30] Distillation is a treatment process that essentially boils produced water to evaporate and then condense the clean water, leaving behind concentrated brine. Reverse osmosis is a filtration process that forces water through a semi-permeable membrane, allowing water to pass through but trapping salt on the other side. Reverse osmosis generally requires a high level of pretreatment to prevent fouling of the membranes, and it is only feasible when salt concentrations in the produced water are less than approximately 25,000 parts per million, according to a study by the Colorado School of Mines.[31] For example, produced water from a gas operation we visited in Wyoming had to first undergo pretreatment to remove solids, hardness, and other contaminants before being put through three stages of reverse osmosis. A third treatment technology, ion exchange, selectively captures sodium ions from produced water and replaces them with others.[32] The water is passed through a large bed of resin beads and sodium ions are adsorbed to (i.e., concentrate on the surface of) the resin. Similar to reverse osmosis, ion exchange faces upper limits on salt concentrations of approximately 7,000 parts per million, according to the Colorado School of Mines.[33] Each of the technologies to remove salt typically generates concentrated brine, which must then be properly disposed of as well.

Cost Is the Primary Factor That Determines How Produced Water Is Managed and Treated

While a variety of factors influence how produced water is managed and the level to which it is treated, cost is the primary factor that oil and gas producers consider when making these decisions. According to producers and agency officials we spoke with, how produced water is managed and treated is primarily an economic decision, made within the bounds of federal and state regulations. In most cases, underground injection is the lowest-cost option and producers we spoke with said that their costs for underground injection range from $0.07 to $1.60 per barrel of produced water.[34] However, if a producer is not operating in close proximity to injection wells, transporting the water via truck or pipeline can significantly increase these costs. Furthermore, producers told us that trucking is one of the most significant cost factors they face, and they seek to minimize this cost by managing the water closer to the production site when possible. For example, according to one producer, trucking costs in Texas range from $0.50 to $1.00 per barrel because injection wells in the area are plentiful, whereas costs in Pennsylvania range from $4.00 to $8.00 per barrel because injection wells are scarce, and the produced water often must be transported out of state. As a result, once trucking is factored in, underground injection may in fact become more costly than other management practices.

Another significant component of cost is whether treatment will be required as part of the management practice being employed. Treatment costs depend heavily on the technologies used, which in turn depend on the quality of the produced water being treated and the level of treatment needed for the disposal or reuse option being considered. For example, there are a number of treatment methods to remove salt from produced water, but each option has a different cost and differing level of effectiveness. Distillation, while effective at removing salts, has significantly higher costs, and representatives from water treatment facilities and producers we spoke to said it can cost from $6.35 to $8.50 per barrel, on average. Reverse osmosis and ion exchange are less costly treatment options for removing salt, but their use is limited by the level of salt content they can remove from produced water, and reverse osmosis can require extensive pretreatment, which can significantly drive up costs. Producers we spoke with who use reverse osmosis and ion exchange to treat produced water told us that their costs range from $0.20 to $0.60 per barrel. In some cases, producers may be able to change the management practice they use to minimize their treatment costs. For example, state regulators told us that when more stringent discharge limits were put into place in Pennsylvania, many shale gas producers in the state stopped discharging produced water to the surface and started to reuse it for hydraulic fracturing because the latter requires a simpler, and less expensive, level of treatment.

In addition to cost, according to our review of the literature and stakeholders we spoke with, produced water management decisions are also influenced by a number of other factors including the following:

- Poor water quality is a key reason most produced water is managed through underground injection, rather than reused or discharged to the surface. However, when water quality is relatively good, as some of it is in the Powder River Basin of Wyoming, management practices such as irrigation and infiltration from surface impoundments may become viable options. Nonetheless, adequate quantities of produced water are needed for irrigation to be a sustainable practice, and the water must be in close proximity to the land it will be used on or producers can face high transportation costs.
- Proximity and region-specific factors, such as geology, also influence which management practices are feasible in a given area. For example, some oil and gas producers are not located in close proximity to injection wells, or the number of available wells is limited by the underlying geology of the area, and therefore producers must manage their produced water some other way. An oil producer we spoke to told us his company would prefer to manage all of its produced water through underground injection for enhanced recovery and disposal. However, opportunities for enhanced recovery at one of the producer's sites are limited by its level of oil production and disposal is constrained by the geology in the area. As a result, the producer told us he manages about 20 percent of the produced water from this site through treatment and discharge, which is significantly more costly and technically challenging than underground injection.
- Climate is also a factor in the decision-making process. Arid climates are favorable for managing produced water using surface impoundments for evaporation, and limited water supplies in certain regions can motivate producers to make the water available for other purposes, such as irrigation.
- Regulatory requirements at the federal or state level can also influence producers' management decisions. As discussed earlier, the changes in discharge limits in Pennsylvania led to a change in management practices by shale gas producers in the state.
- Producers' risk management policies can also influence how they manage the water. For example, regulators we spoke with from California told us that liabilities associated with surface discharges and impoundments are a driving factor in moving away from those practices and toward underground injection. According to these regulators, surface impoundments were commonly used in California

to manage produced water in the past, but in the last few years hundreds of them have closed down and they are no longer widely used.

EPA AND THE STATES WE REVIEWED REGULATE THE MANAGEMENT OF PRODUCED WATER THROUGH A VARIETY OF MEANS

The management of produced water is regulated by EPA and the states we reviewed through a variety of means, depending on how the water is disposed of or reused. EPA regulates the management of produced water that is injected underground under the Safe Drinking Water Act, while it regulates the management of produced water that is discharged into surface waters under the Clean Water Act. Other management practices, such as disposal of the water into surface impoundments, irrigation, and reuse of the water for hydraulic fracturing, are primarily regulated by the state authorities.

Underground Injection Is Regulated under the Safe Drinking Water Act

The management of produced water through underground injection is subject to the Safe Drinking Water Act's Underground Injection Control program. This program is designed to prevent contamination of aquifers that supply, or could supply in the future, public water systems by ensuring the safe operation of injection wells. Under this program, EPA or authorized states generally require producers to obtain permits for their injection wells by, among other things, meeting technical standards for constructing, operating, and testing and monitoring wells.[35] Of the nine states we reviewed, all but Pennsylvania have received approval authority from EPA to implement this program for class II wells, including issuing permits and conducting oversight. In most of these states, the agency that oversees oil and gas activities is responsible for implementing this program. Regardless of whether EPA or the state has authority for implementing the program, EPA regional offices periodically review each state's program and require states to submit an annual report on program activity, according to EPA officials from the regions we spoke with.

As part of the Underground Injection Control program, producers generally must apply for a permit to drill an injection well and supply information, including the location and depth of the proposed well. Furthermore, once EPA or the state has issued an Underground Injection Control permit, producers must observe and record the injection pressure, flow rate, and cumulative volume each month and report this information to the permitting agency annually. In addition, the injection well permit also requires producers to conduct mechanical integrity tests on the wells at least once every 5 years, although EPA and some states require testing to be performed more often, according to officials we spoke with. Officials at many of the state agencies we spoke with said that they observe these tests in person to ensure that the well is mechanically sound. According to officials in each of the eight states we contacted, the state can levy penalties for noncompliance for violations ranging from a failure to submit a report to exceeding the pressure permitted in the well. Enforcement response to noncompliance can range from a warning letter to a fine. EPA can commence a separate action for penalties if it believes that a state's imposition of penalties is insufficient, although EPA officials we spoke with stated that this is rare.

Discharge of Produced Water Is Regulated under the Clean Water Act

The management of produced water through discharge into surface waters is regulated under the Clean Water Act's National Pollutant Discharge Elimination System. Under this program, all facilities that discharge pollutants to surface waters must obtain a permit from EPA or the designated state agency, which is generally the agency responsible for environmental protection or quality. Permits can be tailored to individual facilities or cover multiple facilities within a specific geographic region. To obtain a permit, producers must complete an application that, among other things, describes the waste that will be discharged, where the discharge will take place, and the method of treatment or containment, if applicable. Once the state or EPA has issued a permit, producers must report any discharges, including the volume of effluent and the amount of each pollutant specified in the permit, to the permitting authority at least once per year. EPA has issued regulations establishing Effluent Limitations Guidelines for some onshore oil and gas extraction including shale gas, but these regulations do not apply to coalbed methane extraction.[36]

Of the nine states we reviewed, all but New Mexico have received approval authority from EPA to implement this program for industrial and municipal facilities. EPA requires states with approval authority to submit annual reports on program activity in their state and conduct program reviews every 2 to 5 years. Of these eight states, four—California, Colorado, Pennsylvania, and Wyoming—have issued permits for the discharge of produced water. This is in part because discharging produced water directly from a production site is generally prohibited by regulations implementing the Clean Water Act for locations east of the 98th meridian, which, in the United States, runs from near the eastern border of North Dakota through the eastern portion of Texas, passing near the Dallas-Ft. Worth area. Discharge of produced water from an off-site treatment plant, however, is allowed under the Clean Water Act provided the treated water meets applicable water quality standards, and some states have permitted this activity. For example, a commercial water treatment facility we visited in Pennsylvania treats produced water from shale gas to meet the state's new, more stringent discharge limits and then releases it to a municipal sewer system. In addition, one producer we spoke with in Wyoming told us his company takes some of its produced water to a treatment facility, where it is treated with reverse osmosis and then discharged to a ravine that flows into the Powder River. The eight states that have approval authority to administer the discharge program may levy penalties if they find producers are not complying with their permit, or if they are discharging without a permit, according to officials we spoke with. As with the Underground Injection Control program, EPA may commence a separate action for penalties if it believes a state's penalty determination to be inadequate, but EPA officials we spoke with stated that this is rare.

Other Management Practices Are Regulated by State Authorities

The management of produced water through disposal into a surface impoundment or reuse for irrigation is regulated at the state level in the four states we reviewed where producers employ these practices. For example, the Oil and Gas Conservation Commissions in Colorado and Wyoming are among the state regulatory agencies that allow for disposal in surface impoundments; however, these states have set different standards for the quality of the water that may be placed in the ponds. For example, Colorado generally does not require these ponds to be lined, while Wyoming requires any pond with a total dissolved solids level of 10,000 parts per million or more to be lined. In addition, Wyoming also allows produced water to be used for irrigation or as water for livestock with approval from the Wyoming Department of Environmental Quality.

Some of the states we reviewed also regulate other practices to reuse produced water. For example, regulations in Colorado, Pennsylvania, and Wyoming allow for the application of produced water to roads in certain circumstances. Specifically, Colorado regulations allow produced water to be spread on roads as long as it meets certain requirements and is authorized by the owner of the road. In addition, reuse of produced water for hydraulic fracturing is regulated at the state level for some states.[37] Specifically, some states have regulations that apply to the temporary storage of hydraulic fracturing fluids, including flowback water, on drilling sites. For example, Oklahoma has recently adopted standards for the construction, operation, location, and maintenance of noncommercial ponds used for temporary storage of flowback water. In addition, some states have begun to require producers to disclose the chemical composition of their hydraulic fracturing fluids. Of the nine states we reviewed, four states—Louisiana, Pennsylvania, Texas, and Wyoming—currently require this disclosure.

FEDERAL RESEARCH EFFORTS HAVE FOCUSED ON DESCRIBING THE CHARACTERISTICS OF AND USES FOR PRODUCED WATER, MANAGEMENT OPTIONS, AND TREATMENT METHODS

Over 100 federal research studies conducted during the last 10 years have addressed various aspects of using, managing, and treating produced water. Many federal research projects have focused on describing the characteristics of produced water, such as the volume of water produced from oil and gas activities and the quality of that water. Other research efforts have focused on describing strategies producers could use to manage produced water and the regulatory context for doing so. Federal research also has focused on developing and describing new and existing technologies for treating produced water. Appendix II of this report includes a compilation of the studies we identified.

Federal Research Efforts Have Examined the Characteristics of Produced Water

Several federal agencies, including USGS, the Bureau of Reclamation, a number of DOE national laboratories, and EPA, have issued or sponsored studies describing the characteristics of produced water from oil and gas operations—particularly the volume and quality of produced water. For example, in 2009, USGS published a fact sheet that, among other things, described water disposal issues associated with gas production in the Marcellus Shale.[38] More recently, in 2010, USGS published an article describing the quality of produced water from coalbed methane production, and how untreated or partially treated produced water from these operations may threaten fish and aquatic resources.[39] USGS also maintains a database that provides the location, geologic setting, and chemical composition of produced water samples from locations throughout the United States.

The Department of the Interior's Bureau of Reclamation participated in a study published in 2008 that describes the quantity of produced water generated and specific contaminants it contains from oil and gas production in the western United States.[40] The study was designed to assist producers and others in determining viability of this water for beneficial reuse and for selecting appropriate treatment processes.

Several DOE national laboratories have also sponsored or published research describing the volume and quality of produced water. Argonne National Laboratory has published two studies—a 2004 study that provided basic information about how much produced water is generated and what contaminants are in it,[41] and a 2009 study that describes produced water volumes and management practices in every oil- or gas-producing state in the United States.[42] Oak Ridge National Laboratory has conducted a number of produced water research studies designed to characterize and evaluate the soluble organic compounds that contaminate the water. In addition, NETL cosponsored several such studies, including a 2003 project that provided information about radioactive materials from hydrocarbon production in Mississippi.[43] Another NETL project, completed in 2005, provided a means by which interested stakeholders could access a large quantity of produced water chemistry data for New Mexico oil wells.[44] NETL is currently engaged in two other projects. The first is designed to differentiate produced water from surface water or shallow groundwater in the Marcellus shale area. The other is a multiagency effort with industry, EPA, and USGS, among others, to establish baseline water quality data at a Marcellus drilling site that will be monitored 1 year prior to development and compared to data acquired for 1 year after production begins.

Finally, EPA's Office of Research and Development initiated a study in January 2010 to examine the potential impacts of hydraulic fracturing on drinking water resources and the quality of flowback and produced water. The study plan is currently being reviewed by EPA's Science Advisory Board, and the agency anticipates issuing an interim report on the potential impacts of hydraulic fracturing on drinking water resources in 2012 and a final report in 2014.

Federal Research Has Examined Options for Managing Produced Water and Associated Regulatory Issues

Federal research has also focused on providing information about options producers can use to manage their produced water and the regulations they must follow in doing so. For example, the Energy Policy Act of 2005 mandated that the Department of the Interior, in consultation with EPA, engage the National Academy of Sciences to conduct a study on the effect of coalbed methane production on surface and groundwater resources in selected northern and western states. The study was issued in 2010.[45] Several DOE laboratories have also conducted studies or partnered with industry, universities, and other labs to provide information about managing produced water and associated regulations. For example, in May 2009, NETL, in cooperation with the Ground Water Protection Council, published a report summarizing produced-water-related regulations enacted by states for the purpose of protecting water resources.[46] NETL is also currently partnered with Clemson University and Chevron to study the efficacy of constructing wetlands to provide a low-cost, effective technology for the treatment and potential reuse of produced water. In addition, NETL has partnered with other DOE laboratories to conduct research on management of produced water. For example, from 2001 through 2004, NETL collaborated with Idaho National Laboratory and others to analyze coalbed methane production on an Indian reservation and to evaluate options for managing the associated produced water in an effort to minimize the environmental impacts of the water.

Other DOE national laboratories also have undertaken studies related to the management and regulation of produced water. For example, Argonne National Laboratory published a study in 2002 that described regulatory issues affecting the management of produced water from coalbed methane production.[47] This was followed by a 2004 study that provided information on how produced water is managed and regulated, and the cost of various management practices.[48] More recently, Argonne National Laboratory published a series of studies describing produced water management practices in different energy-producing regions, including the Marcellus formation in the Appalachians and the Fayetteville Shale in Arkansas. In addition, Sandia National Laboratories partnered with a producer to study options for managing produced water from coalbed methane, and published its analysis in 2008.[49] Also, Oak Ridge National Laboratory has developed new approaches for produced water sampling, analysis, and remediation, and Los Alamos National Laboratory is currently conducting research to provide information about how produced water can be used to cultivate algae for biofuel production.

Other federal agencies have also contributed to research on the management and regulation of produced water. Specifically, the Bureau of Reclamation partially funded a publication containing the proceedings from an April 2006 workshop on produced water,[50] and the agency presented information about the beneficial use of produced water at the 2007 International Petroleum Environmental Conference.[51] In addition, in 2006, USGS issued a bibliography of studies from across oil- and gas-producing areas that it had compiled from the last 80 years. These studies describe the effects of produced water on soils, water quality, and ecosystems.[52]

Federal Research Has Explored New Ways and Existing Alternatives for Treating Produced Water

Federal research efforts, primarily conducted by DOE's national laboratories, have also focused on new technologies and treatment methods for produced water. Sandia National Laboratories, for example, has partnered with a producer to conduct pilot testing of a new treatment system to lower the salt content of produced water from coalbed methane sources.[53] NETL has partnered with Los Alamos National Laboratory, the New Mexico Institute of Mining and Technology, and the University of Texas on a long-term project to develop and test a prototype for a new treatment system that uses an innovative filtration method to remove problem contaminants and that would facilitate on-site treatment of produced water. From March 2003 through the end of 2005, NETL also partnered with Oak Ridge National Laboratory and industry to develop and test novel liquid solvents to remove organic substances from produced water. In addition, NETL is currently sponsoring a project to develop high-temperature nanofiltration technology to remove contaminants from produced water. According to the agency, the goal of this project is to minimize environmental impacts from coalbed methane and shale gas operations and allow cost-effective reuse of produced water that will reduce freshwater consumption and disposal costs. More recently, NETL sponsored research that led to the development of a new treatment system that, according to the agency, successfully treated flowback water from a hydraulic fracturing site in Pennsylvania. According to NETL, the treatment system significantly reduced the producer's disposal costs.

Other federal research efforts have been designed to improve existing techniques to treat produced water. For example, NETL partnered with Texas A&M, Argonne National Laboratory, and industry to develop improved reverse osmosis membrane filtration technology for the removal of salt from produced water. The desalination technology developed through this project led to the construction of a large-scale mobile unit and the development of a commercial oilfield treatment system at a site in Texas, according to NETL officials. Similarly, NETL has partnered with industry to develop a process that, when combined with existing reverse osmosis treatment, will facilitate the reuse of produced water by lowering energy requirements needed to treat produced water and by reducing membrane fouling.

AGENCY COMMENTS

We provided a draft of this report to the Department of Energy, the Department of the Interior, and the Environmental Protection Agency for review and comment. None of these agencies provided written comments to include in our report; however, the Environmental Protection Agency and the Department of the Interior provided technical comments, which we incorporated as appropriate.

Sincerely yours,

Anu K. Mittal
Director, Natural Resources and Environment

Frank Rusco
Director, Natural Resources and Environment

Appendix I. Scope and Methodology

Our objectives for this review were to describe (1) what is known about the volume and quality of produced water from oil and gas production; (2) what practices are generally used to manage and treat produced water, and what factors are considered in the selection of each; (3) how the management of produced water is regulated at the federal level and in selected states; and (4) what federal research and development efforts have been undertaken during the last 10 years related to produced water.

To address each of these objectives, we conducted a literature review of studies and other documents on produced water quality and volume, management, and regulation issued by federal agencies and laboratories, state agencies, the oil and gas industry, and academic institutions. These documents included peer-reviewed scientific and industry periodicals, government-sponsored research, and reports from nongovernmental research organizations. We identified this literature through a systematic search of databases such as ProQuest, EconLit, and BioDigest, and used an iterative process to identify the most relevant studies for our review. We believe we have included the key studies and have qualified our findings, where appropriate. However, we may not have identified all of the studies with findings relevant to our objectives. In addition, we reviewed studies that fit the following criteria for selection: (1) the research was of sufficient breadth and depth to provide observations or conclusions directly related to our objectives; (2) the research was targeted specifically toward the volume and quality of produced water, available management practices and treatment methods, regulation of produced water broadly and in selected states, and undertaken by the federal government; and (3) the research was typically published in the last 10 years. We examined key assumptions, methods, and relevant findings of major scientific articles primarily related to water volumes and quality, and treatment methods. Where applicable, we assessed the reliability of the data we obtained and found them to be sufficiently reliable for our purposes.

In addition, we interviewed federal and state regulatory officials; federal scientists from the Environmental Protection Agency's (EPA) Office of Research and Development and the Department of Energy's (DOE) Argonne National Laboratory, Los Alamos National Laboratory, National Energy Technology Laboratory, Oak Ridge National Laboratory, and Sandia National Laboratories; officials from oil and gas exploration and production companies; officials from water treatment facilities; and other experts with experience related to produced water. The federal and state regulatory officials included those with responsibility over oil and gas regulation, as well as clean water and drinking water regulation. We focused our review of management techniques and produced water regulation on nine states—California, Colorado, Kansas, Louisiana, New Mexico, Oklahoma, Pennsylvania, Texas, and Wyoming. We selected eight of these states because the volume of produced water generated within their borders accounts for nearly 90 percent of the produced water generated in the United States as of 2007, the most recent year for which there were available data. In addition, we selected Pennsylvania because of the recent growth in shale gas development in the Marcellus shale formation and the expected potential for large-scale produced water management approaches in this area. While oil shale production has expanded and continues to expand in Texas, North Dakota, and other states, we did not look specifically at produced water from oil shale as part of this review. Furthermore, GAO will be conducting future work on the development of shale gas resources and the use of hydraulic fracturing for oil and gas development and will address these topics more fully in subsequent reports.

We supplemented our literature review and stakeholder discussions with site visits to selected locations in Pennsylvania, Texas, and Wyoming, where we met with oil and gas producers and officials from produced water treatment facilities and discussed issues related to produced water management and treatment and the factors that influence these decisions. We selected these states because of the current and potential volumes of produced water generated, the range of hydrocarbons produced, and the different management and treatment practices employed. We also visited hydraulic fracturing drilling operations, underground injection control sites, and a number of different treatment facilities employing a variety of technologies. To determine what federal research and development efforts have been undertaken during the last 10 years related to produced water, we analyzed information supplied by and conducted interviews with federal officials from DOE and select national laboratories, EPA, and the Department of the Interior's U.S. Geological Survey, Bureau of Land Management, and Bureau of Reclamation.

We conducted this performance audit from October 2010 to January 2012, in accordance with generally accepted government auditing standards. Those standards require that we plan and perform the audit to obtain sufficient, appropriate evidence to provide a reasonable basis for our findings and conclusions based on our audit objectives. We believe that the evidence obtained provides a reasonable basis for our findings and conclusions based on our audit objectives.

Appendix II. List of Ongoing and Completed Federal Produced Water Research Efforts Undertaken during the Last 10 Years

The following is a list of federally sponsored research efforts undertaken in the last 10 years that we identified in consultation with officials from the Department of Energy and select national laboratories, the Department of the Interior, the Environmental Protection Agency, and the National Research Council. These research efforts include those related to the quantity, quality, management, treatment, and use of produced water.

Department of Energy

Argonne National Laboratory

The following studies were published by or prepared for Argonne National Laboratory.

[1] *Analysis of Data from a Downhole Oil/Water Separator Field Trial in East Texas*. February 2001.
[2] Clark, C. E & Veil J. A. (2009). *Produced Water Volumes and Management Practices in the United States*. September.
[3] Harto, Christopher (2011). *Shale Gas – The Energy Water Nexus*. April.
[4] Argonne National Laboratory (2003). *An Introduction to Slurry Injection Technology for Disposal of Drilling Wastes*. September.
[5] Puder, Markus G., Bill Bryson, & John A (2003). *Veil. Compendium of Regulatory Requirements Governing Underground Injection of Drilling Wastes*. February.
[6] Puder, M. G. & Veil J. A. (2006). *Offsite Commercial Disposal of Oil and Gas Exploration and Production Waste: Availability, Options, and Costs*. August.
[7] Veil, John A. & John J. Quinn (2004). *Downhole Separation Technology Performance: Relationship to Geologic Conditions*. November.
[8] Veil, John A. & Maurice B. Dusseault (2003). *Evaluation of Slurry Injection Technology for Management of Drilling Wastes*. May.
[9] Veil, J. A. & M. G. Puder (2006). *Potential Ground Water and Surface Water Impacts from Oil Shale and Tar Sands Energy-Production Operations*. October.

[10] Veil, John A. (2002). *Regulatory Issues Affecting Management of Produced Water from Coal Bed Methane Wells*. February.

[11] Veil, J., Gasper, J., Puder, M. & Leath, P. (2006). *Summary of DOE/PERF Water Program Review*. January.

[12] Veil, John A. (2008). *Thermal Distillation Technology for Management of Produced Water and Frac Flowback Water*. Water Technology Brief #2008-1. May 13.

[13] Veil, J. A. & Quinn, J. J. (2008). *Water Issues Associated with Heavy Oil Production*. November.

[14] Veil, John A. (2011). "Water Management Practices Used by Fayetteville Shale Gas Producers." *Oil & Natural Gas Technology* (June).

[15] Veil, John A. (2010). "Water Management Technologies Used by Marcellus Shale Gas Producers." *Oil & Natural Gas Technology* (July).

[16] Veil, John A., Markus G. Puder, Deborah Elcock, & Robert J. Redweik, Jr (2004). *A White Paper Describing Produced Water from Production of Crude Oil, Natural Gas, and Coal Bed Methane*. January.

Los Alamos National Laboratory

The following studies were published by or prepared for Los Alamos National Laboratory.

[1] Altare, Craig R., Robert S. Bowman, Lynn E. Katz, Kerry A. Kinney, & Enid J. Sullivan (2007). "Regeneration and Long-term Stability of Surfactant Modified Zeolite for Removal of Volatile Organic Compounds from Produced Water." *Microporous and Mesosporous Materials*, *105*, 305-316.

[2] Kwon, Soondong, Enid J. Sullivan, Lynn E. Katz, Robert S. Bowman, & Kerry A. Kinney (2011). "Laboratory and Field Evaluation of a Pretreatment System for Removing Organics from Produced Water." *Water Environment Research*, vol. *83*.

[3] Ranck, J. Michael, Robert S. Bowman, Jeffrey L. Weeber, Lynn E. Katz, & Enid J. Sullivan (2005). "BTEX Removal from Produced Water Using Surfactant-Modified Zeolite." *Journal of Environmental Engineering* (March).

[4] Sullivan, E. J., Dean, C. A., Yoshida, T. M., Cordova, B. Rearick, M., Laur, P. Viszolay, A. Brown, L. & Brown, J. *Chemical Quality Impacts of Oil and Gas Produced Water as a Growth Medium for*

Nannochloropsis Grown at Pilot Scale for Biofuel Production. LA-UR-11-11017.

National Energy Technology Laboratory

The following studies were published by or prepared for the National Energy Technology Laboratory.

[1] Billingsley, R. L. (2005). *Identifying and Remediating High Water Production Problems in Basin-Centered Formations*. December.

[2] Brown, Terry, Carol D. Frost, Thomas D. Hayes, Leo A. Heath, Drew W. Johnson, David A. Lopez, Demian Saffer, Michael A. Urynowicz, John Wheaton, & Mark D. Zoback (2009). *Final Report: Produced Water Management and Beneficial Use*. January.

[3] Burnett, David B. and Mustafa Siddiqui. *Recovery of Fresh Water Resources from Desalination of Brine Produced During Oil and Gas Production Operations*. September 2003–December 2006.

[4] *COAL BED METHANE PRIMER—New Source of Natural Gas – Environmental Implications: Background and Development in the Rocky Mountain West*. February 2004.

[5] *DOE Oil and Natural Gas Water Resources Program*. December 2009.

[6] *Feasibility Study of Expanded Coal Bed Natural Gas Produced Water Management Alternatives in the Wyoming Portion of the Powder River Basin Phase One*. January 2006.

[7] *A Guide to Practical Management of Produced Water from Onshore Oil and Gas Operations in the United States*. October 2006.

[8] *Handbook on Best Management Practices and Mitigation Strategies for Coal Bed Methane in the Montana Portion of the Powder River Basin.* April 2002.

[9] *Handbook on Coal Bed Methane Produced Water: Management and Beneficial Use Alternatives*. July 2003.

[10] *Modern Shale Gas Development in the United States: A Primer*. April 2009.

[11] *Policy Analysis of Produced Water Issues Associated With In-Situ Thermal Technologies*. January 2011.

[12] Remson, Don J. (2005). *Produced Water in the Rocky Mountain Region—Quantity and Quality*. November.

[13] *Review of the U.S. Department of Energy's Environmental Program*. July 1, 2010.

[14] *Siting, Design, Construction and Reclamation Guidebook for Coalbed Natural Gas Impoundments*. May 2006.
[15] Ground Water Protection Council, *State Oil and Natural Gas Regulations Designed to Protect Water Resources*. May 2009.
[16] *Use of Produced Water in Recirculated Cooling Systems at Power Generating Facilities*. September 2006.
[17] Wang, Xixi, Bethany A. Kurz, & Marc D. Kurz (2007). *Subtask 1.18 – A Decision Tool for Watershed-Based Effluent Trading*. February.
[18] Welch, Robert A. and Dwight F. Rychel (2004). *Produced Water from Oil and Gas Operations in the Onshore Lower 48 States*. December.

National Energy Technology Laboratory

The following are past and current projects funded by the National Energy Technology Laboratory, but for which no studies have been published.

[1] *Advanced Membrane Filtration Technology for Cost-Effective Recovery of Fresh Water from Oil and Gas Produced Brine*. Project start: 2003. Project end: 2006.
[2] *Anti-Fouling Reverse Osmosis Desalination System*. Project start: 2009. Project end: 2010.
[3] *Barnett and Appalachian Shale Water Management and Reuse Technologies*. Project start: 2009. Estimated project end: 2011.
[4] *Cleaning Agents for Produced Water Membrane Filters*. Project start: 2004. Project end: 2006.
[5] *Coal Bed Methane Best Management Practices Workshop*. Project start: 2003. Project end: 2004.
[6] *Coalbed Methane Research*. Project start: 2006. Project end: 2008.
[7] *Coalbed Natural Gas Produced-Water Treatment Using Gas Hydrate Formation at the Wellhead*. Project start: 2005. Project end: 2009.
[8] *Coalbed Natural Gas Produced-Water Treatment Using Gas Hydrates*. Project start: 2006. Project end: 2008.
[9] *Coalbed Natural Gas Research*. Project start: 2003. Project end: 2006.
[10] *Comprehensive Lifecycle Planning and Management System for Addressing Water Issues Associated with Shale Gas Development in New York, Pennsylvania and West Virginia*. Project start: 2009. Estimated project end: 2012.
[11] *Cost Effective Recovery of Low-TDS Frac Flowback Water for Re-use*. Project start: 2009. Project end: 2011.

[12] *Cost-Effective Treatment of Produced Water Using Co-Produced Energy Sources for Small Producers*. Project start: 2008. Original project end: 2010 (extended).

[13] *Effects of Irrigating with Treated Oil and Gas Product Water on Crop Biomass and Soil Permeability*. Project start: 2008. Project end: 2010.

[14] *Energy in the Environment-Initiatives 2004-09*. Project start: 2004. Project end: 2009.

[15] *Evaluations of Radionuclides of Uranium, Thorium, and Radium Associated with Produced Fluids, Precipitates, and Sludges from Oil, Gas, and Oilfield Brine Injection Wells in Mississippi*. Project start: 2002. Project end: 2003.

[16] *Field Validation of Toxicity Tests to Evaluate the Potential for Beneficial Use of Produced Water*. Project start: 2004. Project end: 2008.

[17] *GIS and Web-Based Water Resource Geospatial Infrastructure for Oil Shale Development*. Project start: 2008. Estimated project end: 2012.

[18] *Handbooks for Preparing, Evaluation Development, Environmental Plans and Background Development Pertinent to Coal Bed Methane Production*. Project start: 2002. Project end: 2005.

[19] *Hypoxia, Program Review, and Total Petroleum Hydrocarbon Workshop*. Project start: 2006. Project end: 2008.

[20] *Identification, Verification, & Compilation of Produced-Water Best Management Practices for Conventional Oil & Gas Production Operations*. Project start: 2004. Project end: 2007.

[21] *Improving Science-Based Methods for Assessing Risks Attributable to Petroleum Residues in Soil Transferred to Vegetation*. Project start: 2002. Project end: 2005.

[22] *Innovative Water Management Technology to Reduce Environmental Impacts of Produced Water*. Project start: 2008. Estimated project end: 2012.

[23] *An Integrative Framework for the Treatment and Management of Produced Water*. Project start: 2008. Estimated project end: 2011.

[24] *An Integrated Water Treatment Technology Solution for Sustainable Water Resource Management in the Marcellus Shale*. Project start: 2009. Estimated project end: 2011.

[25] *Integration of Water Resource Models with Fayetteville Shale Decision and Support Systems*. Project start: 2009. Estimated project end: 2012.

[26] *Life Cycle Assessment, Produced Water, and Waste Management Analyses*. Project start: 2004. Project end: 2007.

[27] *Long-term field Deployment of a Surfactant Modified Zeolite Vapor Phase Bioreactor System*. Project start: 2004. Project end: 2007.

[28] *Long Term Field Development of a Surfactant-Modified Zeolite/Vapor-Phase Bioreactor System for Treatment of Produced Waters for Power Generation*. Project start: 2004. Project end: 2007.

[29] *Management of Produced Water*. Project start: 2003. Project end: 2006.

[30] *Managing Coalbed Natural Gas Produced Water for Beneficial Uses, Initially Using the San Juan and Raton Basins as a Model*. Project start: 2003. Project end: 2008.

[31] *Membrane Technology for Produced Water at Lea County, NM*. Project start: 2008. Estimated project end: 2011.

[32] *Microbial Ecology of Shale Gas Production Waters*. Project start: 2011. Estimated project end: not established.

[33] *Modified Reverse Osmosis System for Treatment of Produced Water*. Project start: 2000. Project end: 2004.

[34] *Modeling of Water-Soluble Organic Content in Produced Water*. Project start: 2002. Project end: 2005.

[35] *NMWAIDS: A Produced-Water Quality and Infrastruture GIS Database for New Mexico Oil Production*. Project start: 2002. Project end: 2005.

[36] *Northeast National Petroleum Reserve-Alaska Reconnaissance-Level Airborne Contaminants Study*. Project start: 2001. Project end: 2006.

[37] *Northern Cheyenne Indian Reservation (NCIR) Coalbed Natural Gas Resource Assessment and Analysis of Produced-Water Disposal Options*. Project start: 2001. Project end: 2004.

[38] *Novel Cleanup Agents for Membrane Filters Used to Treat Oilfield Produced Water for Beneficial Purposes*. Project start: 2004. Project end: 2007.

[39] *Novel Fouling—Reducing Coatings for Ultrafiltration, Nanofiltration and Reverse Osmosis Membranes*. Project start: 2004. Project end: 2008.

[40] *Pilot Testing: Pretreatment Options to Allow Re-Use of Frac Flowback and Produced Brine for Gas Shale Resource Development*. Project start: 2009. Estimated project end: 2011.

[41] *Pretreatment and Water Management for Frac Water Reuse and Salt Production*. Project start: 2009. Estimated project end: 2011.

[42] *Produced Water Management and Beneficial Use*. Project start: 2005. Project end: 2007.

[43] *Produced Water Management and Beneficial Use/15549 Colorado School of Mines*. Project start: 2005. Project end: 2007.

[44] *Produced Water Management and Beneficial Use/15549 Colorado School of Mines*. (Different portion of the preceding project with distinct project identification.) Project start: 2005. Project end: 2007.

[45] *Produced Water Treatment and Decision Tool*. Project start: 2008. Estimated project end: 2012.

[46] *Provide Support to Produced Water: Osage-Skiatook Petroleum Environmental Research Project*. Project start: 2001. Project end: 2006.

[47] *Range Resources Baseline Monitoring Site for Marcellus Shale Gas*. Project start: 2011. Estimated project end: not established.

[48] *Recovery of More Oil-in-Place at lower Production Costs While Creating a Beneficial Water Resource*. Project start: 2002 Project end: 2006.

[49] *Research and Development Concerning Coalbed Natural Gas—Congressional Mandate*. Project start: 2006. Project end: 2008.

[50] *Research to Enhance Oil and Gas Development and Environmental Protection on Federal Lands: Joint Montana Regional Coalbed Natural Gas Ground-Water Monitoring Program*. Project start: 2005. Project end: 2008.

[51] *Risk Based Data Management System (RBDMS) and Cost Effective Regulatory Approaches (CERA) Related to Hydraulic Fracturing and Geologic Sequestration of CO-2*. Project start: 2009. Estimated project end: 2012.

[52] *Subsurface Drip Irrigation*. Project start: 2007. Estimated project end: 2014.

[53] *Sustainable Management of Flowback Water during Hydraulic Fracturing of Marcellus Shale for Natural Gas Production*. Project start: 2009. Estimated project end: 2012.

[54] *Treating Coalbed Natural Gas Produced Water for Beneficial Use by MFI Zeolite Membranes*. Project start: 2004. Project end: 2008.

[55] *Treatment and Beneficial Reuse of Produced Waters Using a Novel Pervaporation-Based Irrigation Technology*. (NETL in-house project not yet awarded.)

[56] *Treatment of Produced Water by FARADAVIC Electrodialysis and Reverse Osmosis*. Project start: 2009. Project end: 2010.

[57] *Treatment of Produced Waters using a Surfactant Modified Zeolite/Vapor-Phase Bioreactor*. Project start: 2002. Project end: 2006.

[58] *Treatment of Produced Waters Using a Surfactant-Modified Zeolite/Vapor-Phase Bioreactor System*. Project start: 2003. Project end: 2006.

[59] *Treatment of Produced Waters Using a Surfactant-Modified Zeolite/Vapor-Phase Bioreactor System.* (Next phase.) Project start: 2004. Project end: 2006.
[60] *Unconventional High Temperature Nanofiltration for Produced Water Treatment.* (Phase I.) Project start: 2009. Project end: 2010.
[61] *Unconventional High Temperature Nanofiltration for Produced Water Treatment.* (Next phase.) Project start: 2010. Project end: 2012.
[62] *Use of Ionic Liquids in Produced-Water Clean-up.* Project start: 2003. Project end: 2005.
[63] *Use of Stable Isotopes to Discern Marcellus Produced Water When Commingled with Surface Water or Shallow Groundwater.* Project start: 2011. Estimated project end: not established.
[64] *Use of Wetland Plant Species and Communities for Phytoremediation of Coalbed Natural Gas Produced Water and Waters of Quality Similar to that Associated with Coalbed Natutral Gas Deposits of the Powder River Basin.* Project start: 2001. Project end: 2008.
[65] *Using Helicopter Electromagnetic Surveys to Determine the Hydrologic Fate of Coalbed Methane Produced Water.* Project start: 2002. Project end: 2004.
[66] *Water Management Strategies for Improved Coalbed Methane Production in the Black Warrior Basin.* Project start: 2009. Estimated project end: 2012.
[67] *Water-Related Issues Affecting Conventional Oil and Gas Recovery and Potential Oil Shale Development in the Uinta Basin, Utah.* Project start: 2008. Estimated project end: 2011.
[68] *Water & Waste Regulatory Analysis.* Project start: 2006. Project end: 2008.
[69] *Zero Discharge Water Management for Horizontal Shale Gas Well Development.* Project start: 2009. Estimated project end: 2011.

Oak Ridge National Laboratory

The following studies were published by or prepared for Oak Ridge National Laboratory.

[1] Bostick, Debra T., Luo, H. & Hindmarsh, B.(2002). *Characterization of Soluble Organics in Produced Water.* January.
[2] Klasson, K. Thomas, Costas Tsouris, Sandie A. Jones, Michele D. Dinsmore, David W. Depaoli, Angela B. Walker, Sotira Yiacoumi, Viriya Vithayaveroj, Robert M. Counce, & Sharon M. Robinson (2002).

Ozone Treatment of Soluble Organics in Produced Water. Petroleum Environmental Research Forum Project 98-04. January.

[3] McFarlane, J. (2005). "Application of Chemometrics to Modeling Produced Water Contamination." *Separation Science and Technology, 40,* 593-609.

[4] McFarlane, Joanna (2006). *Modeling of Water-Soluble Organic Content in Produced Water.* May.

[5] McFarlane, Joanna (2004). *New Approaches to Produced Water Sampling, Analysis and Remediation at ORNL.*

[6] McFarlane, Joanna (2004). *Measurement, Characterization and Prediction of Organic Solubility in Produced Water*. Presentation at Gas Technology Institute Natural Gas Technologies II Conference and Exhibition, February 8-11.

[7] McFarlane, Joanna (2004). *Offshore Versus Onshore Produced Water Characterization and Models.* Presentation at Gas Technology Institute Natural Gas Technologies II Conference and Exhibition, February 8-11.

[8] McFarlane, Joanna, Debra T. Bostick, & Huimin Luo (2002). *Characterization and Modeling of Produced Water.*

[9] Ren, R. X. (2005). Room Temperature Ionic Liquids for Separating Organics from Produced Water. *Separation Science and Technology*, *40,* 1245-1265.

Sandia National Laboratories

The following study was published by or prepared for Sandia National Laboratories.

[1] Cappelle, Malynda, Randy Everett, William Holub, Richard Kottenstette, & Allan Sattler (2008). *Coal Bed Natural Gas Produced Water Preliminary Pilot Plant Operation and Results*. August.

Department of the Interior

Bureau of Reclamation

The following studies were published by or prepared for the Bureau of Reclamation.

[1] Benko, Katie L. (2009). "Ceramic Membranes for Produced Water Treatment." *World Oil* (April): 1-3

[2] Benko, Katie L & Jörg E. Drewes (2008). "Produced Water in the Western United States: Geographical Distribution, Occurrence and Composition." *Environmental Engineering Science*, vol. *25*, no. 2, 239-246.
[3] Benko, Katie and Jörg Drewes, Pei Xu, & Tzahi Cath (2009). "Use of Ceramic Membranes for Produced Water Treatment." *World Oil*, Gulf Publishing Company, vol. *230*, no. 4 (April).
[4] Drewes, Jörg E, Pei Xu, Dean Heil, &Gary Wang (2009). *Multibeneficial Use of Produced Water Through High-Pressure Membrane Treatment and Capacitive Deionization Technology*. Desalination and Water Purification Research and Development Program Report No. *133*. February.
[5] Dundorf, Steve & Katie Benko. (2007) "Geographical Assessment of Potential for Beneficial Use of Produced Water." Presentation at International Petroleum Environmental Conference. November.
[6] Colorado Waters Resources Research Institute, Colorado State University, *Produced Water Workshop*, (April 4-5, 2006). The publication of workshop results was partially funded by the Bureau of Reclamation.

U.S. Geological Survey

The following studies were published by or prepared for the U.S. Geological Survey.

[1] Engle, Mark A., Carleton R. Bern, Richard W. Healy, James I. Sams, John W. Zupancic, & Karl T. Schroeder (2011). "Tracking solutes and water from subsurface drip irrigation application of coalbed-methane produced waters, Powder River Basin, Wyoming." *Environmental Geosciences*, v. *18*, no. 3 (September), 1-19.
[2] Farag, Aida M., David D. Harper, Anna Senecal, & Wayne A. Hubert (2010). "Potential Effects of Coalbed Natural Gas Development on Fish and Aquatic Resources." Chapter 11 in *Coalbed Natural Gas: Energy and Environment*. Nova Science Publishers, Inc.
[3] Healy, Richard W., Cynthia A. Rice, Timothy T. Bartos, & Michael P. McKinley (2008). "Infiltration from an impoundment for coal-bed natural gas, Powder River Basin, Wyoming: Evolution of water and sediment chemistry." Water Resources Research, Vol. *44*, W06424, June.

[4] Healy, Richard W., Timothy T. Bartos, Cynthia A. Rice, Michael P. McKinley, & Bruce D. Smith (2011). "Groundwater chemistry near an impoundment for produced water, Powder River Basin, Wyoming USA." *Journal of Hydrology, 403,* 37-48.

[5] Kharaka, Y.K., & J.K. Otton, 2003, Environmental Impacts of Petroleum Production: Initial Results from the Osage-Skiatook Petroleum Environmental Research Sites, Osage County, Oklahoma: USGS Water-Resources Investigations Report 03-4260.

[6] Orem, William H., Calin A. Tatu, Harry E. Lerch, Cynthia A. Rice, Timothy T. Bartos, Anne L. Bates, Susan Tewalt, & Margo D. Corum (2007). "Organic compounds in produced waters from coalbed natural gas wells in the Powder River Basin, Wyoming, USA." Applied Geochemistry *22* (May), 2240-2256.

[7] Otton, James K. *Environmental Aspects of Produced-water Salt Releases in Onshore and Coastal Petroleum-producing Areas of the Conterminous U.S. – A Bibliography.* Open-File Report 2006-1154.

[8] Peterman, Zell E., Joanna N. Thamke, Kiyoto Futa, & Thomas A. Oliver. *Strontium Isotope Detection of Brine Contamination in the East Poplar Oil Field, Montana.* Open-File Report 2010-1326.

[9] Rice, Cynthia A., Timothy T. Bartos, and Margaret S. Ellis. *Chemical & Isotopic Composition of Water in the Fort Union and Wasatch Formations of the Powder River Basin, Wyoming and Montana: Implications for Coalbed Methane Development.* Coalbed Methane of North America – II, 2002: 53-70.

[10] Rowan, E. L., Engle, M. A., Kirby, C. S. & Kraemer, T. F. (2011). *Radium Content of Oil- and Gas-Field Produced Waters in the Northern Appalachian Basin: Summary and Discussion of Data.* Scientific Investigations Report 2011-5135.

[11] Smith, Bruce D., Joanna N. Thamke, Michael J. Cain, Christa Tyrrell, & Patrica L. Hill (2004). *Helicopter Electromagnetic and Magnetic Survey Maps and Data, East Poplar Oil Field Area, Fort Peck Indian Reservation, Northeastern Montana, August.* Open-File Report 2006-1216, Version 1.0.

[12] Soeder, Daniel J. & William H. Kappel (2009). *Water Resources and Natural Gas Production from the Marcellus Shale.* USGS Fact Sheet 2009-3032. May.

Environmental Protection Agency

[1] Environmental Protection Agency, *Draft Hydraulic Fracturing Study Plan* (April 28, 2011).

National Research Council

Committee on Management and Effects of Coalbed Methane Development and Produced Water in the Western United States, Committee on Earth Resources, Board on Earth Sciences and Resources, Water Science and Technology Board, Division on Earth and Life Studies, National Research Council of the National Academies. *Management and Effects of Coalbed Methane Produced Water in the Western United States.* 2010.

End Notes

[1] Most of the balance of the nation's energy is supplied by coal, nuclear power, and hydropower and renewable resources.

[2] Ceramic beads are sometimes used in lieu of sand when hydraulically fracturing a well. Both are used to prevent the fractures from closing when the injection has stopped.

[3] GAO, *Energy-Water Nexus: Improvements to Federal Water Use Data Would Increase Understanding of Trends in Power Plant Water Use*, GAO-10-23 (Washington, D.C.: Oct. 16, 2009); *Energy-Water Nexus: Many Uncertainties Remain about National and Regional Effects of Increased Biofuel Production on Water Resources*, GAO-10-116 (Washington, D.C.: Nov. 30, 2009); *Energy-Water Nexus: A Better and Coordinated Understanding of Water Resources Could Help Mitigate the Impacts of Potential Oil Shale Development*, GAO-11-35 (Washington, D.C.: Oct. 29, 2010); *Energy-Water Nexus: Amount of Energy Needed to Supply, Use, and Treat Water Is Location-Specific and Can Be Reduced by Certain Technologies and Approaches*, GAO-11-225 (Washington, D.C.: Mar. 23, 2011).

[4] The federal government issues leases for federal lands and waters to oil and gas operators who in turn pay royalties to the government on the oil and gas they produce. These royalty payments totaled approximately $9 billion in 2009.

[5] Proven, or proved, reserves are defined as oil and gas that geological and engineering data demonstrate with reasonable certainty to be recoverable in future years from known reservoirs under existing economic and operating conditions. While the United States also has extensive offshore oil and gas reserves, the focus of this report is limited to onshore oil and gas production.

[6] This report includes oil and gas currently produced onshore in the United States. It excludes some kinds of oil production from shale formations that are not generally being produced commercially at the current time, as well as oil sands, which are primarily produced in Canada.

[7] Viscosity is a measure of the resistance of a fluid to flow.

[8] Pub. L. No. 93-523 (1974), codified as amended at 42 U.S.C. §§ 300f–300j-26.

[9] The other classes of Underground Injection Control program wells are as follows: class I wells are used for the disposal of hazardous and certain nonhazardous waste; class III wells are used to inject fluids for mineral extraction; class IV wells are used to dispose of hazardous or radioactive wastes, into or above an underground source of drinking water; class V wells are used to dispose of other nonhazardous wastes; and class VI wells are used for carbon sequestration. Class IV wells are currently banned.

[10] The Federal Water Pollution Control Act Amendments of 1972, Pub. L. No. 92-500, § 2, 86 Stat. 816, codified as amended at 33 U.S.C. § 1251 et seq. (commonly referred to as the Clean Water Act).

[11] Surface waters refer to navigable waters, tributaries to navigable waters, interstate waters, the oceans out to 200 miles, and intrastate waters that are used by interstate travelers, for recreation or other purposes, as a source of fish or shellfish sold in interstate commerce, or for industrial purposes by industries engaged in interstate commerce. Point sources are wastes discharged from discrete sources such as pipes. The Clean Water Act defines point source as "any discernible, confined and discrete conveyance, including but not limited to any pipe, ditch, channel, tunnel, conduit, well, discrete fissure…or vessel or other floating craft from which pollutants are or may be discharged."

[12] Argonne National Laboratory, *Produced Water Volumes and Management Practices in the United States*, ANL/EVS/R-09/1 (Argonne, III.: September 2009).

[13] An aquifer is a natural underground layer, often of sand or gravel, that contains water.

[14] In this example, conventional gas well is used to refer to formations typically consisting of porous sandstone that can be produced using traditional drilling methods.

[15] An individual shale gas well is typically fractured between 10 and 16 times.

[16] Argonne National Laboratory, *Produced Water Volumes and Management Practices in the United States.*

[17] Argonne National Laboratory reported that more than 98 percent of produced water generated from onshore oil and gas wells in 2007 was injected underground. EPA officials we spoke with estimated that over 90 percent of produced water generated from onshore wells is still injected underground but could not provide a precise figure for current volumes.

[18] Although approximately 80 percent of these injection wells are used for enhanced recovery and the remaining 20 percent are used for disposal, only about 59 percent of produced water is injected into these wells for enhanced recovery, and about 40 percent is injected for disposal.

[19] According to EPA officials, there are currently only six active injection wells for produced water disposal in Pennsylvania. As a result, producers that want to dispose of produced water through underground injection would generally have to transport the water to authorized injection wells in Ohio or West Virginia, and trucking can be expensive. However, EPA officials we interviewed said that in the past 2 years producers have shown interest in drilling additional injection wells for disposal in Pennsylvania, and EPA has received permit applications for new wells.

[20] Other less commonly used management practices, such as reusing the produced water in industrial settings for dust control or cooling water, may also be options for producers.

[21] Argonne National Laboratory, *Produced Water Volumes and Management Practices in the United States.*

[22] There are limited exceptions to this prohibition. For example, produced water from onshore oil and gas production sites located west of the 98th meridian that will be put to beneficial use for agriculture or wildlife watering can be discharged directly from the production site.

[23] Pennsylvania regulatory officials told us that almost all conventional oil and gas producers use surface discharges to manage produced water, following treatment; however, almost all shale gas producers in Pennsylvania stopped using this practice after the state's surface discharge standards were made more stringent in 2011.

[24] Surface impoundments are also used for the temporary storage of produced water prior to managing it in some other way.

[25] National Research Council, *Management and Effects of Coalbed Methane Produced Water in the Western United States* (Washington, D.C.: 2010).

[26] National Research Council, *Management and Effects of Coalbed Methane Produced Water in the Western United States*.

[27] Managed irrigation combines irrigation using produced water with preventative or intervention soil management, such as the addition of gypsum and elemental sulfur to improve the soil structure. Managed irrigation is necessary to prevent substantial deterioration of the soil structure caused by the salts and sodium present in produced water.

[28] Ground Water Protection Council and ALL Consulting, *Modern Shale Gas Development in the United States: A Primer* (Oklahoma City, Okla.: April 2009).

[29] pH is a measure of how acidic or basic the water is. Excessively high or low pH can be detrimental for the use of water.

[30] The ratio of sodium to calcium and magnesium ions in produced water is an important property affecting the infiltration and permeability of the soil. The sodium adsorption ratio is an index used to measure the hazard related to sodium abundance in the soil.

[31] Colorado School of Mines, *An Integrated Framework for Treatment and Management of Produced Water* (Golden, Colo.: November 2009). The Colorado School of Mines also reported that total dissolved solids (i.e., salt concentrations) in produced water from conventional oil and gas in the western United States range from 1,000 to 400,000 parts per million. Producers we interviewed in other parts of the United States told us total dissolved solids in their produced water range from 40,000 to 300,000 parts per million.

[32] Ion exchange can also be used to remove hardness.

[33] Colorado School of Mines, *An Integrated Framework for Treatment and Management of Produced Water*.

[34] The costs presented in this report generally do not include construction costs for the injection well or treatment facility.

[35] Certain existing injection wells were authorized by rule if the well was properly inventoried within 1 year after the effective date of the applicable Underground Injection Control program.

[36] On October 20, 2011, EPA announced it is developing standards for wastewater discharges produced by natural gas extraction from underground coalbed and shale formations. No comprehensive set of national standards exists at this time for the disposal of wastewater discharged from natural gas extraction activities. The agency plans to propose wastewater rules for coalbed methane in 2013 and for shale gas in 2014.

[37] In general, the process of hydraulic fracturing is not directly regulated at the federal level at this time. Congress exempted most hydraulic fracturing activities from EPA's jurisdiction in the 2005 Energy Policy Act amendments to the Safe Drinking Water Act; however, the agency has authority to regulate hydraulic fracturing when diesel fuels are used in fracturing fluids or propping agents. EPA is currently studying the impacts of hydraulic fracturing and plans to issue an interim report on the potential impacts of hydraulic fracturing on drinking water resources in 2012 and a final report in 2014.

[38] Daniel J. Soeder and William M. Kappel, *Water Resources and Natural Gas Production from the Marcellus Shale*, USGS Fact Sheet 2009-3032 (May 2009).

[39] Aida M. Farag, David D. Harper, Anna Senecal, and Wayne A. Hubert, "Potential Effects of Coalbed Natural Gas Development on Fish and Aquatic Resources," chap. 11 in *Coalbed Natural Gas: Energy and Environment* (Nova Science Publishers, Inc., 2010).

[40] Katie L. Benko and Jörg E. Drewes, "Produced Water in the Western United States: Geographical Distribution, Occurrence, and Composition," *Environmental Engineering Science*, vol. 25, no. 2 (2008):.239 – 246.

[41] John A. Veil, Markus G. Puder, Deborah Elcock, and Robert J. Redweik, Jr., Argonne National Laboratory, *A White Paper Describing Produced Water from Production of Crude Oil, Natural Gas, and Coal Bed Methane* (January 2004).

[42] Argonne National Laboratory, *Produced Water Volumes and Management Practices in the United States.*

[43] *Evaluations of Radionuclides of Uranium, Thorium, and Radium Associated with Produced Fluids, Precipitates, and Sludges from Oil, Gas, and Oilfield Brine Injection Wells in Mississippi*, a study cosponsored by the National Energy Technology Laboratory, 2002-2003.

[44] *NM WAIDS: A Produced-Water Quality and Infrastructure GIS Database for New Mexico Oil Production*, a study sponsored by the National Energy Technology Laboratory, project start: 2002, project end: 2005.

[45] National Research Council, *Management and Effects of Coalbed Methane Produced Water in the Western United States.*

[46] Ground Water Protection Council, *State Oil and Natural Gas Regulations Designed to Protect Water Resources*, a report funded by the Department of Energy and prepared for the National Energy Technology Laboratory, May 2009.

[47] John A. Veil, Argonne National Laboratory, *Regulatory Issues Affecting Management of Produced Water from Coal Bed Methane Wells* (February 2002).

[48] John A. Veil et al., Argonne National Laboratory, *A White Paper Describing Produced Water from Production of Crude Oil, Natural Gas, and Coal Bed Methane.*

[49] Malynda Cappelle, Randy Everett, William Holub, Richard Kottenstette, and Allan Sattler, Sandia National Laboratories, *Coal Bed Natural Gas Produced Water Preliminary Pilot Plant Operation and Results* (August 2008).

[50] Colorado Waters Resources Research Institute, Colorado State University, *Produced Water Workshop* (April 4-5, 2006).

[51] Steve Dundorf and Katie Benko, *Geographical Assessment of Potential for Beneficial Use of Produced Water* (presented at the International Petroleum Environmental Conference, Houston, 2007).

[52] James K. Otton, Department of the Interior, U.S. Geological Survey, *Environmental Aspects of Produced-water Salt Releases in Onshore and Coastal Petroleum-producing Areas of the Conterminous U.S. – A Bibliography*, Open-File Report 2006-1154.

[53] Malynda Cappelle et al., Sandia National Laboratories, *Coal Bed Natural Gas Produced Water Preliminary Pilot Plant Operation and Results.*

In: Further Explorations in the Energy-Water Nexus ISBN 978-1-62081-318-8
Editors: D. A. Hughes and B. F. Smith

Chapter 2

PRODUCED WATER VOLUMES AND MANAGEMENT PRACTICES IN THE UNITED STATES*

C. E. Clark and J. A. Veil

EXECUTIVE SUMMARY

Produced water is water from underground formations that is brought to the surface during oil or gas production. Produced water is the largest volume by-product or waste stream associated with oil and gas exploration and production. The cost of managing such a large volume of water is a key consideration to oil and gas producers. Previous national produced water volume estimates are in the range of 15 to 20 billion barrels (bbl; 1 bbl = 42 U.S. gallons) generated each year in the United States. This is equivalent to a volume of 1.7 to 2.3 billion gallons per day. In comparison, the Washington D.C. government and surrounding jurisdictions provide about 300 million gallons per day of drinking water to local residents, businesses, and other users. This represents only about 13% of the daily produced water volume. As another example, consider a backyard swimming pool that is 20 ft wide by 50 ft long and 5 ft deep. The volume of water needed to fill such a pool is about 37,000 gal or about 900 bbl.

* This is an edited, reformatted and augmented version of the U.S. Department of Energy, Office of Fossil Energy, National Energy Technology Laboratory, dated September 2009.

Produced water volume generation and management in the United States are not well characterized at a national level. The U.S. Department of Energy asked Argonne National Laboratory to compile data on produced water associated with oil and gas production to better understand the production volumes and management of this water.

This report provides a current estimate for the volume of produced water generated from oil and gas production in the United States. The volume estimate represents a compilation of data obtained from numerous state oil and gas agencies and several federal agencies for 2007, where possible. The total volume of produced water estimated for 2007 is about 21 billion bbl. This equals an average of 57.4 million bbl/day. Produced water is generated from most of the nearly 1 million actively producing oil and gas wells in the United States.

Argonne contacted state oil and gas agencies in the 31 states with active oil and gas production to obtain detailed information on produced water volumes and management. Not all states had readily available precise produced water volume figures. In a few states, the agencies had very complete data records easily obtainable from online sources. Other states had summary-level volume data without much detail or had data available only in in-house data repositories. The most challenging states were those that had no produced water data at all. In those cases, we calculated estimates through extrapolation and correlations using hydrocarbon production and produced water volumes from neighboring states. To obtain federal data, Argonne contacted the Minerals Management Service, the Bureau of Land Management, and the U.S. Environmental Protection Agency.

State and federal onshore production contribute the majority of produced water (more than 20 billion bbl) in the United States. A significant amount of produced water (more than 700 million bbl or about 3% of the national total) is also generated from federal offshore production activities and from production on tribal lands. The five states with greatest produced water volumes in 2007 were Texas, California, Wyoming, Oklahoma, and Kansas. The produced water volumes from these states represent nearly 75% of total U.S. production (onshore and offshore), as shown in Figure 1. Texas, with more than 7.3 billion bbl, contributed 35% of the total volume of produced water generated in the United States in 2007. The contributions from the other five states with produced water volumes exceeding 1 billion bbl were California (12%), Wyoming (11%), Oklahoma (11%), Kansas (6%), and Louisiana (5%).

The greatest produced water contributors are not necessarily the greatest producers of oil and gas. While Texas was the largest gas producer in the

United States (nearly 6,900,000 million cubic feet [Mmcf] in 2007), federal offshore production activities provided the largest volume of crude of more than 467,000,000 bbl. Although federal offshore production generates nearly 27% of U.S. crude oil production, less than 3% of total U.S. produced water is generated from federal offshore activities.

In addition to total volumes produced, it is useful to consider the water-to-oil ratios (WORs) and water-to-gas ratios (WGRs) from production activities as this information can be used to evaluate the relative production age of resources within the production lifetime. To that end, we asked the agencies to provide produced water volumes by hydrocarbon types (i.e., crude oil, conventional gas, coal bed methane, unconventional gas, or other), to the extent the data were available at that level of detail. Most states were unable to break out produced water volumes for all categories, but some states could at least provide estimates of produced water from oil production vs. natural gas production. States that segregated produced water by hydrocarbon type categorized 6,666,144,270 bbl of produced water. Eighty-seven percent (5,770,327,439 bbl) of produced water came from oil production activities.

We also were able to calculate separate WORs and WGRs for several states that reported produced water separately for hydrocarbon type (WORs for 14 states and WGRs for 11 states). A national average WOR calculated using the production-weighted ratios from the 14 states was 7.6 bbl/bbl. When offshore production was added to the onshore production, the total average U.S. WOR was 5.3 bbl/bbl. A national average WGR calculated using the production-weighted ratios from the 11 states was 260 bbl/Mmcf. When offshore production was added to the onshore production, the total average U.S. WGR was 182 bbl/Mmcf.

Two general water management themes were followed by most U.S. operators in 2007. More than 98% of produced water from onshore wells is injected underground. Approximately 59% is injected into producing formations to maintain formation pressure and increase the output of production wells. Another 40% of produced water from onshore wells is injected into nonproducing formations for disposal. More than 91% of offshore produced water, including the water from inshore platforms in Cook Inlet, Alaska, is discharged to the ocean. Most of the remaining volume is reinjected for enhanced recovery. The remaining 2% of national produced water volume was managed through evaporation ponds, offsite commercial disposal, beneficial reuse, and other management methods.

The produced water volumes WORs, and WGRs included in this report are based on the best data available. The best data was far from complete, however. Argonne needed to extrapolate information from states with detailed produced water data for other nearby states with limited produced water information. The national average WOR and WGR values were estimated using data from less than half of the oil and gas producing states. Several of the states that have large numbers of producing wells, particularly in mature fields with many stripper wells (e.g., Texas, Oklahoma) did not have produced water data segregated by production type. If those data had been included with the states already in the average, it is probable that the WOR and WGR would be substantially larger.

1. Introduction

1.1. Purpose

Produced water volume generation and management in the United States are not well characterized at a national level. The U.S. Department of Energy (DOE) asked Argonne National Laboratory to compile data on produced water associated with oil and gas production to better understand the production volumes and management of this water. The purpose of this report is to improve understanding of produced water by providing detailed information on the volume of produced water generated in the United States and the ways in which produced water is disposed or reused. As the demand for fresh water resources increases, with no concomitant increase in surface or ground water supplies, alternate water sources, like produced water, may play an important role.

1.2. Background

Produced water is water from underground formations that is brought to the surface during oil or gas production. Because the water has been in contact with hydrocarbon-bearing formations, it contains some of the chemical characteristics of the formations and the hydrocarbons. It may include water from the reservoir, water previously injected into the formation, and any chemicals added during the production processes. The physical and chemical properties of produced water vary considerably depending on the geographic location of the field, the geologic formation, and the type of hydrocarbon product being produced. Produced water properties and volume also vary throughout the lifetime of a reservoir.

Produced water is the largest volume by-product or waste stream associated with oil and gas exploration and production. Previous national produced water volume estimates are in the range of 15 to 20 billion barrels (bbl; 1 bbl = 42 U.S. gallons) generated each year in the United States (API 1988, 2000; Veil et al. 2004). However, the details on generation and management of produced water are not well understood on a national scale.

1.3. Overview

Argonne National Laboratory developed detailed national-level information on the volume of produced water generated in the United States and the manner in which produced water is managed. This report presents an overview of produced water, summarizes the study, and presents results from the study at both the national level and the state level. Chapter 2 presents background information on produced water, describing its chemical and physical characteristics, where it is produced, and the potential impacts of produced water to the environment and to oil and gas operations. A review of relevant literature is also included. Chapter 3 describes the methods used to collect information, including outreach efforts to state oil and gas agencies and related federal programs. Because of the inconsistency in the level of detail provided by various state agencies, the approaches and assumptions used to extrapolate data values are also discussed. In Chapter 4, the data are presented, and national trends and observations are discussed. Chapter 5 presents detailed results for each state, while Chapter 6 presents results from federal sources for oil and gas production (i.e., offshore, onshore, and tribal lands). Chapter 7 summarizes the study and presents conclusions.

2. Produced Water

This chapter provides a brief introduction to produced water and produced water management. A more thorough discussion of produced water can be found in *A White Paper Describing Produced Water from Production of Crude Oil, Natural Gas, and Coal Bed Methane* (Veil et al. 2004).

2.1. Definition of Produced Water

Produced water is water found in the same formations as oil and gas. When the oil and gas are produced to the surface, the produced water is brought to the surface, too. It is also referred to as "brine," "saltwater," or "formation water." Produced water contains some of the chemical characteristics of the formation from which it was produced and associated hydrocarbons.

Produced water properties (both physical and chemical) and volume vary considerably depending on the geographic location of the field, the geologic formation, the type of hydrocarbon product being produced, and the lifetime of a reservoir. For example, early in the life of an oil well, oil production is high and water production is low. As the production age of the well increases, the oil production decreases and the water production increases. When the cost of managing produced water exceeds the profit from selling oil, production is terminated and the well is closed. This is contrary to the typical production cycle of a coal bed methane (CBM) well. Initially CBM wells produce large volumes of water, which decline over time. Methane production is initially low, increases over time to a peak, and then decreases.

2.2. Previous Produced Water Volume Estimates

While one of the purposes of this report is to present a current estimate of produced water volumes, it is useful to know previous estimates and the assumptions used in arriving at those estimates. Khatib and Verbeek (2003) estimated a global average of 210 million bbl of water produced each day, which resulted in an annual estimate for 1999 of 77 billion bbl of produced water. It is not clear how those authors derived their estimate (collecting and compiling accurate produced water data within a single country is a challenging task, as described in Chapter 3). International estimates must be taken as approximations. U.S. onshore estimates of produced water from oil and gas activities were estimated at 21 billion bbl in 1985 and 18 billion bbl in 1995 by the American Petroleum Institute (API 1988, 2000) and 14 billion bbl in 2002 by Veil et al. (2004). Significant additional volumes of produced water are generated at U.S. offshore wells.

The volume of water produced from oil and gas wells does not remain constant over time. The water-to-hydrocarbon ratio increases over the life of the well. Initially, water represents a small percentage of produced fluids. Over time, the percentage of water increases and the percentage of hydrocarbon decreases. In the study by Khatib and Verbeek (2003), a world average estimate was reported to be 3 bbl of water for each barrel of oil. U.S. wells are typically further in their production lifetime than the global average. Veil et al (2004) reported an average of more than 9.5 bbl of water for each barrel of oil in the United States.

CBM wells, in contrast to oil and gas wells, produce a large volume of water early in their life, and the water volume declines over time. CBM wells have increased considerably since the 1995 API study year and were not included in the study by Veil et al. (2004). As a result, the actual produced water volume in the United States is most likely higher than the 14 to 18 billion bbl estimates previously reported.

2.3. Characteristics of Produced Water

The physical and chemical properties of produced water vary considerably depending on the geographic location of the field, the geologic formation from where the water was produced, and the type of hydrocarbon product being produced. For those sites where waterflooding is conducted, the properties and volumes of the produced water may vary dramatically due to the injection of additional water into the formation to increase hydrocarbon production. The major constituents of concern are salt content (often expressed as salinity, conductivity, or total dissolved solids [TDS]), oil and grease (various organic compounds associated with hydrocarbons in the formation), inorganic and organic compounds introduced as chemical additives to improve drilling and production operations, and naturally occurring radioactive material.

Understanding the constituents of specific produced waters aids regulatory compliance and the selection of appropriate management options for the produced water, such as secondary recovery and disposal. Oil and grease is the most important constituent in offshore produced water. It is also an important one for onshore produced water. Note that the term oil and grease refers not a single chemical, but rather to an analytical test that measures the presence of many families of organic chemicals. A study of produced water in the western United States found the oil and grease content to range from 40 mg/L to 2,000 mg/L (Benko and Drewes 2008). Another important constituent of concern in onshore operations is the salt content of produced water. According to Cline (1998), most produced waters are more saline than seawater. Benko and Drewes (2008) found the TDS concentration of produced water in the western

United States to vary between 1,000 mg/L and 400,000 mg/L. While high TDS can increase maintenance costs, detecting TDS assists in defining the pay zones of a formation, when coupled with resistivity measurements (Breit et al. 1998).

Produced water from oil production activities often contains constituents in addition to those that are naturally found within the formation. Additional water is often needed to maintain sufficient pressure in a reservoir for oil production. Produced water may be reused for this purpose, but the water may also be supplied from additional sources including groundwater and seawater. These additional water sources may contain additional solids and microorganisms (Chapelle 2001; Dowd et al. 2000). To combat scaling and maintain production efficiency, chemical additives such as corrosion and scale inhibitors, emulsion breakers, coagulants, and solvents may be used in drilling operations, production operations, and separations processing. The production of a well can be improved by utilizing the appropriate scale inhibitor and well-treatment chemicals according to the characteristics of the formation (Breit et al. 1998). However, these additives can become part of the produced water and can affect its overall toxicity.

Produced water from gas production has different characteristics than produced water from oil production. In addition to formation water, water produced from gas production will contain condensed water, which is water that was in the vapor phase while in the reservoir but then condenses into a liquid state in the production separation system.

Produced water from CBM production differs from produced water from both oil and gas production. Oil and grease are less of a concern from CBM water than other produced waters. To recover the methane in CBM reservoirs, the hydrostatic pressure that caused the adsorption of methane to the coal bed is reduced through the removal of water from the reservoir via CBM wells. Characteristics of CBM water that may affect reuse are salinity, sodicity, and to a lesser extent iron, manganese, and boron (ALL 2003).

2.4. Produced Water Management

While produced water can be reused if certain water quality conditions are met, most produced water generated is disposed. For offshore production activities, produced water is usually disposed of through direct ocean discharge after treatment. For onshore production activities, produced water is managed in a variety of ways. According to API (2000), 92% of the 18 billion barrels of produced water generated in 1995 was managed through injection. Three percent of the 18 billion bbl of produced water was discharged under National Pollutant Discharge Elimination System (NPDES) permits; nearly all of this water was generated from coal bed methane operations. Two percent was managed through beneficial reuse. The remaining 3% was disposed through other methods including evaporation, percolation pits, and publicly owned treatment works (API 2000). These management methods are briefly described in this section.

2.4.1. Discharge

For U.S. offshore operations, the majority of produced water is discharged to the ocean and is subject to applicable regulatory requirements. Offshore produced water discharges are authorized by NPDES general permits issued by U.S. Environmental Protection Agency (U.S. EPA) regional offices. All of the permits contain a monthly average limit of 29 mg/L and a 42 mg/L daily maximum limit for oil and grease.

In the U.S., wells in "coastal" areas generally may not discharge produced water, although there are exceptions for some wells that are located in Alaska's Cook Inlet (Veil et al. 2004). Offshore activities around the globe typically discharge to the ocean, although countries have different discharge standards.

Discharge activities from onshore production are primarily limited to water produced from CBM activities or from other oil and gas wells in the western United States that have produced water with low salinity. Currently, the decision to allow CBM water discharges is made by either state agencies or U.S. EPA regional offices, depending on the state's permitting authority (Veil 2002). U.S. EPA has initiated a study on CBM to determine if national effluent limitation guidelines are needed.[1]

2.4.2. Underground Injection for Disposal

Twenty-one percent of produced water generated in the United States in 1995 was managed through injection wells. Under U.S. EPA rules, produced water injection wells are classified as Class II wells and managed by the Underground Injection Control program. Class II wells may be classified as II-R (enhanced recovery), II-D (disposal), or II-H (hydrocarbon storage) (U.S. EPA 2008a). Class II injection wells are designed and constructed to inject fluids to authorized zones in a manner that does not allow migration into underground sources of drinking water (USDWs). Injection wells for disposal are often located in formations that enable water to enter at pressures below the fracture pressure and are isolated from USDWs and hydrocarbon-producing formations.

In addition to locating a formation with appropriate characteristics to receive the water, it is also important that the produced water is chemically compatible with the receiving formations. This may require treatment prior to injection to manage excessive solids, dissolved oil, corrosion, chemical reactions, or microorganisms.[2]

2.4.3. Underground Injection for Increasing Oil Recovery

Most produced water generated onshore is injected to maintain reservoir pressure and hydraulically drive oil toward a producing well. This practice is often referred to as enhanced oil recovery (EOR), and is also referred to as water flooding or steam flooding, depending upon the temperature of the water. In 1995, injection for EOR managed 71% of the produced water generated (API 2000). When managed via injection for EOR, produced water becomes a resource rather than a waste product.[3]

2.4.4. Evaporation

While not practiced as widely as injection, evaporation is a simple management strategy for produced water. A typical evaporation approach directs produced water into a pond with a large surface area. Water then passively evaporates from the surface. The rate of evaporation will depend on the size, depth, and location of the pond, as well as the quality of the produced water. In semiarid regions, hot, dry air moving from a land surface will result in high evaporation rates for smaller ponds. As the concentration of solids and other constituents increase in the remaining produced water, the evaporation rate will decrease over time. This management approach can be conducted onsite in small evaporation ponds or can be conducted offsite at large commercial facilities. One drawback to evaporation ponds is that they are often attractive to waterfowl. To mitigate the risk of exposure to hydrocarbons and other compounds within the produced water, netting is often applied over the ponds.[4]

2.4.5. Offsite Commercial Disposal

When onsite management is not practical, operators may send their produced water offsite to a commercial disposal facility. Typically, produced water is removed from well locations periodically and transported via truck to an offsite facility.5 Offsite commercial disposal is often chosen by small producers who find this option more feasible than constructing, operating, and closing onsite facilities, or by operators who do not have access to suitable formations for produced water injection (Puder and Veil 2006).

2.4.6. Beneficial Reuse

There are several management strategies in addition to enhanced oil recovery that use produced water for beneficial purposes. These primarily include agricultural and industrial uses.

Beneficial reuse of produced water for agricultural purposes includes reuse for crop irrigation, livestock watering, wildlife watering and habitat, aquaculture, and hydroponic vegetable cultivation. However, reuse of produced water represents just a small fraction of total agricultural withdrawals. In 2000, the total U.S. water withdrawals (both surface water and groundwater withdrawals) for irrigation was 1,190 billion barrels and for livestock water and aquaculture was 47.[5] billion barrels (Hutson et al. 2004). A significant challenge to using produced water for agricultural purposes is the salinity of the water. Crops vary in their susceptibility to salinity, and when salinity rises above a threshold for a species, the crop yield will decrease. Another concern for using produced water is the sodicity, which refers to the amount of sodium in the water.[6] Johnston et al. (2008) reported that produced water from coal bed natural gas activities in the Powder River Basin increased the sodium ion concentration within the soil profile when applied to mixed-hay cropland. Excessive levels of sodium can lead to loss of soil structure, reducing the hydraulic conductivity of soils and creating conditions that may limit or prevent plant growth. When sodic soils are wet, they become swollen and waterlogged. When these soils are dry, the surface becomes crusted and resistant to water infiltration or plant emergence. Additional elements in produced water can cause harm to plants when present in sufficient quantities. ALL (2003) notes that plant toxicity can be caused by elevated concentrations of chloride, sodium, and boron.

Reuses of produced water for industrial purposes include water for hydraulic fracturing at oil and gas sites, water for power generation, dust control, and fire control. To initiate production operations and enhance ongoing production, wells may be hydraulically fractured. Fracturing requires hundreds of thousands of barrels of water. In locations where local water sources are insufficient or otherwise unavailable, produced water may be used for fracturing. An additional water source in these operations is the flowback water from a previous fracturing activity. The feasibility of using produced water as cooling water for power generation was considered in a research project funded by the U.S. Department of Energy and the Electric Power

Research Institute (DeFilippo 2004). The economics in the evaluated case did not appear favorable, although there is at least one case of produced water being used as boiler feed water at a cogeneration plant (Brost 2002).

3. APPROACH

To better understand the volume of produced water generated and how it is managed in the United States, Argonne contacted state oil and gas agencies in the 31 states with active oil and gas production to obtain detailed information on produced water volumes and management. As a first step, Argonne contacted the state oil and gas directors to introduce the project, briefly explain the scope of the work, and ask for a contact who could provide detailed information for that state. State agencies were selected due to their long-term direct experience with oil and gas activities in the specific state and the data management systems that most states employ for tracking production data. An example letter is provided in Appendix A of this report.

3.1. Information Request and Questionnaire

Once an appropriate contact person was identified for each state oil and gas agency, two questionnaires were sent to the contact. The first questionnaire targeted the produced water study outlined in this report and is described below. The second questionnaire was for a similar study on flowback water from hydraulic fracturing activities. The results of that study are discussed in a separate report.

The produced water questionnaire was divided into two parts, produced water volume and produced water management. The two tables from the questionnaire, along with instructions, are shown below.

Part I – Produced Water Volume

1. Please provide information on the volume of produced water generated in your state. Where available, please enter the actual volume on a yearly basis for 2007 or the next most recent year into Table 1 or indicate how we can access your electronic data management system. Even if you don't have information on the volume generated, but you do have information on the volume

reinjected (assuming that most produced water from your state is reinjected), that is valuable information too, and should be entered in Table 1. To the extent possible, we would like to see the produced water volume estimates broken down by the type of hydrocarbon produced by the well as shown in Table 1. If you do not have quantitative information on the volume of produced water generated, please give us your educated "best estimate" of the volume.

2. If you do not have quantitative information on produced water generation or reinjection volumes, please provide information on the annual volume of each type of hydrocarbon produced in your state for 2007 or the next most recent year. This information should be entered into the last column of Table 1.
3. Please provide an estimate of the ratio of produced water to each type of hydrocarbon (for example, 7 bbl of water per 1 bbl of oil).

Part II – Produced Water Management

4. Please provide information on how produced water is disposed or otherwise managed in your state. Where available, please enter the number of wells that manage produced water by each of the management practices on a yearly basis for 2007 or the next most recent year into Table 2. If you do not have quantitative information on produced water management practices, please give us your educated "best estimate" of the percentage of wells following each management practice.

Table 1. Produced Water Volume Information

Type of Hydrocarbon	# Wells Producing That Type of Hydrocarbon	Total Volume of Produced Water (bbl/year)	Volume of Hydrocarbon Produced (bbl/year or Mmcf/year)	Ratio of Water to Hydrocarbon (bbl/bbl or bbl/Mmcf)
Crude oil				
Conventional gas				
Coal bed methane				
Unconventional gas				
Other				
Total				

Table 2. Produced Water Management Information

Management Practice	# Wells Using That Practice	Total Volume of Produced Water Managed by That Practice (bbl/year)	Percentage of Produced Water Managed by That Practice
Injection for enhanced recovery			
Injection for disposal			
Surface discharge			
Evaporation			
Offsite commercial disposal			
Beneficial reuse			
Other			

5. For any produced water entered under the beneficial reuse or other categories, please provide, to the extent possible, more details on the actual methods employed.
6. If another state agency has responsibility for regulating or overseeing some of the produced water management practices in your state, please indicate the name of the agency and a contact person.

The information requested through the questionnaire represented Argonne's "wish list." We did not realistically expect that all or even most states would be able to provide that level of detail or complete all of the boxes in the tables. The responses from the states (described in Chapter 5) matched those expectations. In Table 2, the intent of "the number of wells using this practice" column was to learn how many oil and gas wells direct produced water using the various management approaches. This information was not provided consistently. As these terms were not defined, many states interpreted the "number of wells using that practice" to mean the number of injection wells, discharges, evaporating ponds, disposal sites, and reuse sites managing the water.

3.2. Additional Production Information

In order to account for produced water generated from wells outside of the scope of state oil and gas agencies, efforts to obtain production information at the federal level were also undertaken. For offshore production activities, the questionnaire described above was sent to the U.S. Department of the Interior's Minerals Management Service (MMS), Gulf of Mexico Regional Office. Oil and gas production information for 2007 was also obtained from the Minerals Revenue Management Program of the MMS for federal onshore and tribal production activities. The oil and gas production estimates from these federal resources as well as the responses from state agencies were compared with available production data from the U.S. Department of Energy's Energy Information Administration (EIA) to identify any inconsistencies and to assist in the estimation of produced water volume from production activities where data were not available.

Several oil and gas platforms operate in the waters of Cook Inlet, Alaska. Although Cook Inlet is considered coastal rather than offshore waters, U.S. EPA's national discharge standards allow discharge to Cook Inlet waters. At the time this report was prepared, Alaska was going through a process to gain the authority to administer the NPDES program but had not yet obtained that authority. The current active NPDES general permit for Cook Inlet discharges was issued by U.S. EPA Region 10. Therefore, Argonne contacted the Region 10 office to obtain information on the discharge volumes for the Cook Inlet platforms.

4. Analysis and Results

4.1. Response to Questionnaire

The produced water questionnaire was sent to 31 state oil and gas agencies. We received information on produced water generation or management from 28 states. For those states that did not directly provide the requested information, efforts were made to extract available data from accessible reports from oil and gas agency websites. Links to these sites are available at the Produced Water Management Information System website.[7] The questionnaire was also sent to and completed by the MMS to capture offshore production activities and management practices.

4.2. Incomplete Data

A major challenge in this study was dealing with the incompleteness of available data, as not all states track produced water data. For those states where produced water generation volume was unknown, produced water management data was used. When neither produced water management information nor production information were available, estimates of produced water volume were determined from hydrocarbon production information. Production information was obtained from state oil and gas agencies and federal resources as detailed in Chapter 5.

While information on produced water generation and management was obtained for most states, the detail of the information varied widely. Additionally, produced water data were unavailable for two states: Kentucky and Tennessee. The water-to-oil ratio (WOR) and water-to-gas ratio (WGR) for states without available produced water generation volumes were assumed to equal the corresponding ratios of adjacent states. For those states where the surrounding states had limited produced water information, the ratios were taken from several adjacent or neighboring states. This procedure is described in the state summaries in Chapter 5.

Produced water volumes from oil and gas production activities on tribal lands were not readily obtainable. Because obtained production information only provided total hydrocarbon production data from tribal lands, the U.S. average produced water-to-hydrocarbon ratio was used to estimate produced water volumes from these tribal lands.

The summary tables in this chapter address the incompleteness of the data by indicating where the data were obtained and the assumptions used in extrapolating data from other sources. The following categories appear under "Data Source" in the tables:

1. Provided directly to Argonne by state agency;
2. Obtained via published report;
3. Obtained via electronic database;
4. Obtained from website in form other than a published report or electronic database;
5. Obtained from the EIA; and
6. Estimated produced water generation volumes from hydrocarbon production volumes.

While the primary method for obtaining information was through questionnaires and direct communication with state agencies, multiple sources were often used to compile production information for each state.

Table 3. U.S. Onshore and Offshore Oil, Gas, and Produced Water Generation for 2007

State	Crude Oil (bbl/year)	Total Gas (Mmcf)	Produced Water (bbl/year)	Data Source[a]
Alabama	5,028,000	285,000	119,004,000	1
Alaska	263,595,000	3,498,000	801,336,000	1
Arizona	43,000	1,000	68,000	1, 2
Arkansas	6,103,000	272,000	166,011,000	2, 3
California	244,000,000	312,000	2,552,194,000	2, 3
Colorado	2,375,000	1,288,000	383,846,000	1,3
Florida	2,078,000	2,000	50,296,000	1
Illinois	3,202,000	No data	136,872,000	1, 5
Indiana	1,727,000	4,000	40,200,000	1, 2
Kansas	36,612,000	371,000	1,244,329,000	1, 2
Kentucky	3,572,000	95,000	24,607,000	1, 3, 6
Louisiana	52,495,000	1,382,000	1,149,643,000	1
Michigan	5,180,000	168,000	114,580,000	1, 3
Mississippi	20,027,000	97,000	330,730,000	1
Missouri	80,000	No data	1,613,000	1
Montana	34,749,000	95,000	182,266,000	1
Nebraska	2,335,000	1,000	49,312,000	1
Nevada	408,000	0	6,785,000	1, 2
New Mexico	59,138,000	1,526,000	665,685,000	1
New York	378,000	55,000	649,000	2
North Dakota	44,543,000	71,000	134,991,000	2, 4
Ohio	5,422,000	86,000	6,940,000	1, 2
Oklahoma	60,760,000	1,643,000	2,195,180,000	2, 6
Pennsylvania	1,537,000	172,000	3,912,000	3
South Dakota	1,665,000	12,000	4,186,000	1, 2
Tennessee	350,000	1,000	2,263,000	4, 6
Texas	342,087,000	6,878,000	7,376,913,000	3, 4
Utah	19,520,000	385,000	148,579,000	1
Virginia	19,000	112,000	1,562,000	1, 4
West Virginia	679,000	225,000	8,337,000	1
Wyoming	54,052,000	2,253,000	2,355,671,000	1, 4
State Total	**1,273,759,000**	**21,290,000**	**20,258,560,000**	
Federal Offshore	467,180,000	2,787,000	587,353,000	1
Tribal Lands	9,513,000	297,000	149,261,000	2, 6
Federal Total	**476,693,000**	**3,084,000**	**736,614,000**	
U.S. Total	**1,750,452,000**	**24,374,000**	**20,995,174,000**	

[a] 1 = provided directly to Argonne by state agency; 2 = obtained via published report or electronically; 3 = obtained via electronic database; 4 = obtained from website in form other than a published report or electronic database; 5 = obtained from EIA; 6 = produced water volumes are estimated from production volumes.

4.3. Inconsistent Data

Hydrocarbon production information can be assumed to be accurate, due to the value of the oil and gas and the various reporting requirements for production. State production volumes provided to Argonne were compared with production estimates reported by the DOE's EIA. While there is some variation in reported production volumes, this is generally explained in two ways. Production volumes reported in state databases may not be the final reported volumes for a specific year. An official state report on production has final production volumes, and, when available, this source was used for production information. In most instances, final reported volumes agreed with production volumes reported by the EIA. When differences did exist, they were often due to the type of information reported. Most states reported produced hydrocarbon volumes, although some states provided saleable hydrocarbon volumes. Produced hydrocarbon volumes were preferentially used to determine the WORs and WGRs.

While produced water volumes are not required for reporting, many states do keep track of this information. The accuracy of the reported volumes depends upon the methods used by the producer. For instance, producers in Florida do not measure volumes produced, but estimate produced water volumes according to hydrocarbon production. Many states could provide only produced water generation estimates from water injection volumes (e.g., Louisiana). The water injected into Class II wells, especially injection wells for enhanced recovery, may include sources of water other than produced water (e.g., surface water, sea water), which could overestimate the total volume of produced water generated. Additionally, those states where produced water can be managed in ways other than injection would not capture those management strategies, resulting in underreporting total produced water generation. While the authors recognize the uncertainty, a method of accounting was difficult to achieve for the variability in the data collection methods, as the details of the collection methods are unknown. The summary statistics in this chapter have been rounded to the nearest 1,000 unit (e.g., bbl for oil and water, million cubic feet [Mmcf] for gas). Chapter 5 provides detailed information on the type of information obtained as well as the reported numbers (not rounded) for each state.

4.4. Production Summary

In 2007, U.S. onshore and offshore oil and gas production activities generated nearly 21 billion barrels of produced water. Table 3 provides production information for each state. State production totals include production from federal lands within each state.

State and federal onshore production contribute the majority of produced water (more than 20 billion barrels) in the United States. A significant amount of produced water (more than 700 million bbl) is also generated from federal offshore production activities and from production on tribal lands. The five states with greatest produced water volumes in 2007 were Texas, California, Wyoming, Oklahoma, and Kansas. The produced water volumes from these states represent nearly 75% of total U.S. production (onshore and offshore) as shown in Figure 1. Texas alone contributed 35% of the total volume of produced water generated in the United States in 2007.

The greatest produced water contributors are not necessarily the greatest producers of oil and gas. While Texas was the largest gas producer in the United States (nearly 6,900,000 Mmcf in 2007), federal offshore production activities provided the largest volume of crude oil of more than 467,000,000 bbl (see Figure 2). Although federal offshore production generates nearly 27% of U.S. crude oil production, less than 3% of total U.S. produced water is generated from federal offshore activities. While Alaska and the federal offshore were two of the top three producers of oil and gas in 2007, they are ranked 7th and 9th in produced water volume generation.

In addition to total volumes produced, it is useful to consider the WOR and WGR from production activities. This information can be used to evaluate the relative production age of resources within the production lifetime. Table 4 lists water-to-hydrocarbon ratios from states where produced water data could be provided according to the predominant hydrocarbon produced at a specific location.

For WORs, the states values ranged from 2.5 bbl/bbl for South Dakota to 42.7 bbl/bbl for Illinois. The offshore ratio was even smaller at 1.04 bbl/bbl. For WGRs, the state values ranged from 0.04 bbl/Mmcf in South Dakota to more than 1,200 bbl/Mmcf for Kansas. The offshore ratio was 86 bbl/Mmcf.

A national average WOR and WGR can be estimated by developing a production-weighted average of the individual state values. Each state's WOR is multiplied by the oil production, and each WGR is multiplied by the gas production from gas-only wells to give water production from oil and gas production. These water volumes are summed, then they are divided by the sum of the oil and gas production values to estimate the WOR and WGR.

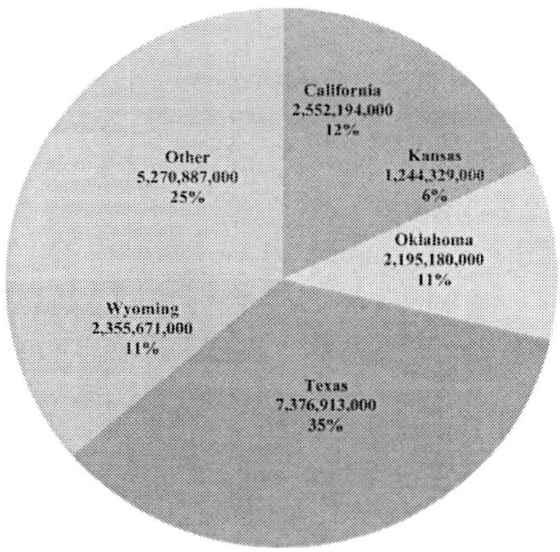

Figure 1. Total U.S. produced water generated (barrels) in 2007 by state (five states with the greatest produced water generation are shown).

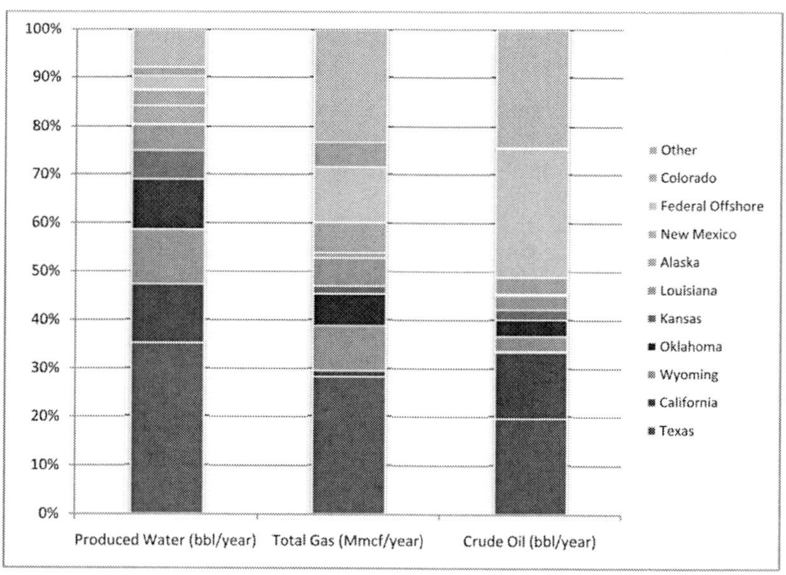

Figure 2. Ten largest produced water generators and their contributions to gas and crude oil production.

The national average onshore WOR was 7.6 bbl/bbl. When offshore production was added to the onshore production, the total average U.S. WOR was 5.3 bbl/bbl. A national average onshore WGR was 260 bbl/Mmcf. When offshore production was added to the onshore production, the total average U.S. WGR was 182 bbl/Mmcf.

Table 4. Water-to-Hydrocarbon Ratio for 2007 from Available Data[a]

State	Water to Crude Oil (bbl/bbl)	Water to Gas (bbl/Mmcf)
Alabama	7.7	282
Alaska	2.9	4.4
California	10.5	7.6
Florida	24.2	Not available
Illinois	42.7	Not available
Kansas	21.8	1208
Mississippi	13.5	35.9
Missouri	20.3	Not available
Montana	4.0	453
Nebraska	20.9	358
Nevada	16.6	Not available
New Mexico	9.0	91.5
North Dakota	3.0	18.0
South Dakota	2.5	0.0
Virginia	Not available	17.7
Onshore Ratio[b]	**7.6**	**260**
Federal Offshore	1.04	86.0
Total Ratio[b]	**5.3**	**182**

[a] States not shown in Table 4 did not distinguish produced water volumes by hydrocarbon type; therefore, the WORs and WGRs could not be calculated.

[b] Onshore and total ratios were determined using the total volumes in each category (water, oil, and gas).

The WOR from onshore activities determined in this study of 7.57 is within the range of previous reports of 7 bbl (Lee et al. 2002) and 9.5 bbl (Veil et al. 2004) of water per bbl of oil. The U.S. ratio for 2007 reveals that 88% of the material brought to the surface for oil production is water. According to Weideman (1996), water can comprise as much as 98% of the material brought to the surface for crude oil wells nearing the end of their productive lives. While cessation of operations depends on the point when managing the produced water is no longer profitable, the WOR determined in this study suggests that production of available resources is mature for most onshore operations in the United States. In the fluids brought to the surface from federal offshore oil wells, water comprises 51% on average.

The authors caution that the national average WOR and WGR values estimated in this study are likely to underestimate the true values. The averages are based on only 14 states (WOR) and 11 states (WGR). Many of the states that have large numbers of producing wells, particularly in mature fields with many stripper wells (e.g., Texas, Oklahoma), did not have produced water data segregated by production type. If those data had been included with the data from states already in the average, it is probable that the national average WOR and WGR values would be substantially larger.

4.5. Produced Water Management Summary

Management information was obtained for nearly 17.1 billion bbl (81%) of the 20.9 billion bbl of produced water generated in 2007. The vast majority of produced water in the United States, 95.2% of the reported volume, was managed through injection. More than half of the produced water (55.4%, or 8.6 billion bbl) was injected for enhanced recovery in 2007. More than one-third (38.9%, or 6.0 billion bbl) of produced water was injected for disposal. Surface discharges managed 4.4% (700,000,000 bbl) of the total reported volume of produced water managed in 2007. Table 5 provides summary management information about injection and surface discharges for each state and the federal offshore area where such information was available. Those states that did not provide produced water management information have only reported total generation volumes. The remaining produced water volume was managed through evaporation ponds, offsite commercial disposal, beneficial reuse, and other management methods. Detailed management information is available for each state in Chapter 5.

Table 5 indicates that most of the produced water managed through injection (both for enhanced recovery and disposal) is generated through onshore production activities. While injection activities take place in the federal offshore area, the total injection volume is small compared to the total volume of produced water generated in the federal offshore area. Table 5 also reveals that some states reported greater volumes of produced water managed than generated for 2007 (e.g., Mississippi, Utah). Injection for enhanced recovery often includes makeup water from various sources, which could explain the discrepancies.

Table 5. U.S. Produced Water Volume by Management Practice for 2007

State	Volume of Produced Water (bbl/year)				
	Injection for Enhanced Recovery	Injection or Disposal	Surface Discharge	Total Managed	Total Generated
Alabama	7,500,000	33,000,000	78,000,000	119,000,000	119,004,000
Alaska	1,037,909,000	39,914,000	35,480,000	1,113,302,000	801,336,000
Arizona	No data	35,000	No data	35,000	68,000
Arkansas	45,489,000	120,169,000	No data	166,011,000	166,011,000
California	1,764,609,000	558,188,000	No data	2,322,797,000	2,552,194,000
Colorado	No data	No data	No data	No data	383,846,000
Florida	34,762,000	15,534,000	No data	50,296,000	50,296,000
Illinois	135,264,000	No data	No data	135,264,000	136,872,000
Indiana	34,500,000	5,700,000	No data	40,200,000	40,200,000
Kansas	444,319,000	800,009,000	No data	1,244,329,000	1,244,329,000
Kentucky	No data	No data	No data	No data	24,607,000
Louisiana	66,261,000	1,034,092,000	No data	1,149,643,000	1,149,643,000
Michigan	25,000,000	90,000,000	No data	115,000,000	114,580,000
Mississippi	389,614,000	281,563,000	No data	671,177,000	330,730,000
Missouri	No data	1,611,000	No data	1,613,000	1,613,000
Montana	109,217,000	46,807,000	No data	182,266,000	182,266,000
Nebraska	31,588,000	14,337,000	500,000	49,312,000	49,312,000
Nevada	No data	No data	No data	No data	6,785,000
New Mexico	449,489,000	348,142,000	No data	797,630,000	665,685,000

Table 5. (Continued)

State	Volume of Produced Water (bbl/year)				
	Injection for Enhanced Recovery	Injection or Disposal	Surface Discharge	Total Managed	Total Generated
New York	No data	No data	No data	No data	649,000
North Dakota	64,873,000	65,321,000	No data	134,991,000	134,991,000
Ohio	487,000	6,137,000	No data	6,940,000	6,940,000
Oklahoma	940,272,000	1,254,132,000	No data	2,195,180,000	2,195,180,000
Pennsylvania	No data	No data	No data	No data	3,912,000
South Dakota	2,122,000	1,853,000	85,000	4,146,000	4,186,000
Tennessee	No data	No data	No data	No data	2,263,000
Texas	5,011,062,000	2,365,476,000	No data	7,376,913,000	7,376,913,000
Utah	78,251,000	62,051,000	21,080,000	173,145,000	148,579,000
Virginia	No data	No data	No data	No data	1,562,000
West Virginia	3,942,000	No data	3,857,000	8,337,000	8,337,000
Wyoming	No data	No data	No data	No data	2,355,671,000
State Total	**10,676,530,000**	**7,144,071,000**	**139,002,000**	**18,057,527,000**	**20,258,560,000**
Federal Offshore	48,673,000	1,298,000	537,381,000	587,353,000	587,353,000
Tribal Lands	No data	No data	No data	No data	149,261,000
Total	**10,725,203,000**	**7,145,369,000**	**676,383,000**	**18,644,880,000**	**20,995,174,000**

While surface discharge is used to manage some onshore produced water, the vast majority of surface discharges occur in the ocean. In 2007, more than 676,000,000 bbl of produced water were discharged to water bodies. Figure 3 shows the sources by percent of produced water surface discharges in the United States. The federal offshore managed the greatest volume of produced water through surface discharge. Including discharges to the Cook Inlet in Alaska, nearly 85% of the volume of produced water managed through surface discharge (573,000,000 bbl) is directed to oceans. The remaining 15% of produced water that is managed through surface discharge is from onshore production activities.

Alabama also manages a significant amount of produced water from coal bed methane activities through surface discharge. As produced water from coal bed methane production is often of a higher quality than other than produced water from more conventional production sources, some states including Alabama allow surface discharge after treatment.

Beneficial reuse is currently not a significant nationwide practice for produced water management. In 2007, 1,338,000 bbl were reported as managed through beneficial use. Reported volumes were limited to Oklahoma, South Dakota, and Utah, although other states including New York and Montana reported practices without providing detailed data on the volumes managed through beneficial reuse. It is likely that some other states have modest amounts of beneficial reuse of produced water, particularly for CBM production when the water is relatively fresh prior to treatment.

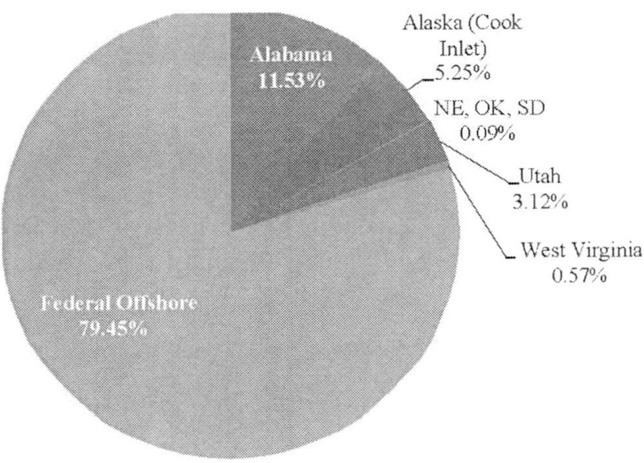

Figure 3. Sources by percent of total U.S. surface discharges for produced water in 2007 (total discharges amounted to more than 676,000,000 bbl in 2007).

5. State-by-State Summary

5.1. Alabama

The State Oil and Gas Board of Alabama provided produced water generation and management data (Bolin 2008). As of December 2007, Alabama had 6,544 wells producing hydrocarbons, with the majority of these wells producing CBM (5,209 wells). The remaining wells produce crude oil (848 wells), conventional gas (475 wells), and unconventional gas (12 wells). Alabama specified shale gas for unconventional gas. From the 6,544 wells, 119,003,506 bbl of produced water were generated in 2007. CBM and crude oil production were responsible for the majority of the produced water generation, with each producing 78,087,301 and 38,495,257 bbl, respectively. Produced water generation from conventional gas was 2,420,635 bbl, and unconventional gas production generated 313 barrels of produced water.

Hydrocarbon production for 2007 included crude oil (5,028,4285 bbl), CBM (114,981 Mmcf), conventional gas (170,049 Mmcf), and unconventional gas (52 Mmcf). Conventional gas wells also produced 2,091,226 bbl of condensate. From these production volumes, the following ratios were determined: WOR of 7.66 bbl/bbl, WGR of 679 bbl/Mmcf for CBM, WGR of 14.23 bbl/Mmcf for conventional gas, and WGR of 6.0 bbl/Mmcf for unconventional gas.

The majority of generated produced water in Alabama is managed through surface discharges for CBM, which require NPDES permits. The Alabama Department of Environmental Management (ADEM) administers the NPDES permit program. Permitted surface discharge accounts for 78,000,000 bbl of produced water from CBM production (5,204 wells). Injection for disposal manages 33,000,000 bbl of produced water per year for 1,174 wells. A smaller amount of produced water is injected for enhanced recovery with 7,500,000 bbl managed for that purpose. The remaining produced water (approximately 500,000 bbl) is managed through offsite commercial disposal. There is no beneficial reuse of produced water in Alabama.

5.2. Alaska

The Alaska Oil and Gas Commission (OGC) supplied both produced water generation information and management information (McMains et al.

2008). In 2007, Alaska had 1,875 wells with the majority of these wells producing crude oil (1,715 wells). The remaining wells produced conventional gas (160 wells). Crude oil production volumes were 263,595,162 bbl in 2007 and generated 765,121,124 bbl of produced water. Conventional gas production generated 735,412 bbl of produced water in the same period to produce 166,261 Mmcf of gas. These volumes result in a WOR of 2.9 bbl/bbl of crude oil and WGR of 4.4 bbl/Mmcf of conventional gas. Total gas production (including gas coproduced from oil wells) was 3,498,322 Mmcf as reported in the Alaska OGC production data archives (Alaska OGC 2009).

Injection for enhanced recovery manages 94.8 percent of produced water in Alaska. Injection for enhanced recovery occurs at 842 wells, and these wells manage 1,037,908,618 bbl of water. This volume exceeds the total volume of available produced water, due to the addition of seawater injected for enhanced recovery. The remaining produced water (39,913,859 bbl) is managed by 30 injection wells for disposal. Management information for discharges from production activities in the Cook Inlet was obtained from U.S. EPA Region 10. In 2007, 35,479,866 bbl of produced water were directly discharged into the Cook Inlet (Shaw 2009). Discharges in the Cook Inlet operate under the U.S. EPA NPDES permit program until October 31, 2011, when discharge activities associated with the oil and gas industry will be transferred to the Alaska NPDES program (Hartig 2009).

5.3. Arizona

Annual oil, water, and gas production volumes are available in the December monthly production reports on the Arizona Oil and Gas Conservation Commission's website at http://www.azogcc.az.gov/. In 2007, 42,692 bbl of oil, 67,682 bbl of water, and 654.530 Mmcf of gas were produced. Most produced water generated in Arizona is managed through injection for disposal. Some produced water is trucked to an approved disposal facility in New Mexico. There is no injection for enhanced recovery at this time. The Arizona Geological Survey, which provides administrative and staff support to the Arizona Oil and Gas Conservation Commission, provided a spreadsheet of Class II injection wells, which reported that 35,116 bbl of produced water were injected for disposal in 2007 (Rauzi 2008). Historic hydrocarbon production volumes are available in the report, *Annual Oil, Gas, and Helium Production in Arizona 1954-2008* (Rauzi 2009).

5.4. Arkansas

Production information was obtained through the Arkansas Oil and Gas Commission's (Arkansas OGC's) annual production report (Arkansas OGC 2007). For 2007, total production volumes in the state were 6,102,538 bbl of crude oil and 271,846 Mmcf of gas. While the report provides production information for South Arkansas and North Arkansas, specific information on the type of produced gas (i.e., conventional, unconventional, or CBM) was not provided.

The Arkansas OGC also provided produced water management information (Pearson 2009). While there is significant water produced from bromine production activities, and the Arkansas OGC tracks the volumes produced and managed that are associated with bromine production, such water is managed through Class V wells and is not included in the management information here. There are 106 active Class II injection wells for enhanced oil recovery and 448 Class II injection wells that dispose of produced water. Most produced water was managed through injection for disposal, with 120,169,316 bbl injected. Injection for enhanced recovery managed 45,488,886 bbl of produced water. An additional 352,997 bbl of produced water was injected, but the type of injection well was not specified. The total injected volume of produced water for 2007 was 166,011,199 bbl.

5.5. California

Produced water and hydrocarbon production information were obtained from the California Department of Conservation's *2007 Preliminary Report of California Oil and Gas Production Statistics* (CDOC 2008a). Oil production in 2007 was reported to be 244,000,000 bbl. The report separates natural gas production according to production associated with oil zones (associated production) and production not associated with oil zones (nonassociated production). Associated production was estimated to total 219,000 Mmcf, and nonassociated production was estimated at 93,200 Mmcf. Rounding to significant figures, the total gas production was 312,000 Mmcf. The water produced from all activities was 2,552,194,253 bbl, which accounts for both onshore and state and federal offshore production. The total produced water volume oil and associated gas production was 2,549,823,222 bbl in 2007 (WOR: 10.45 bbl/bbl). The produced water from nonassociated gas production was 2,162,592 bbl in 2007 (WGR: 6.93 bbl/Mmcf).

Produced water management information was not provided, although some information could be determined from an online database (CDOC 2008b). The online database captures underground injection volumes and reported 2,322,797,059 bbl of water were injected. In 2007, 558,187,847 bbl were injected for disposal, and 1,764,609,212 bbl were injected for enhanced recovery. These volumes may include ocean water or other brine. This does not account for all of the produced water produced in California, but it does suggest that much of the water is injected for the purposes of enhancing oil and gas production activities. Produced water that is not injected for enhanced recovery or as water disposal is treated for beneficial use or is disposed of through settling ponds and ocean outfall.

5.6. Colorado

The Colorado Oil and Gas Conservation Commission (COGCC) publishes production data in an online database, which was used to obtain produced water information for 2007 (COGCC 2009). For that year, crude oil production was 2,374,921 bbl. Total gas production was 1,288,109.976 Mmcf, with CBM contributing 478,434.041 Mmcf to the gas total. Produced water volumes totaled 383,845,756 bbl. Detailed information on produced water management was not available, although it is generally managed through evaporation ponds or injection (Kerr 2009).

5.7. Florida

The Bureau of Mining and Minerals Regulation provided produced water generation and management information for the State of Florida (Taylor 2008). All wells in Florida produce oil, with most producing both oil and gas. There are 56 oil producing wells in the state, and 52 of those wells produce both oil and gas. Crude oil production was 2,077,773 bbl in 2007, and 2,010.932 Mmcf of natural gas was produced during the same period. Total produced water volume from the 56 wells was 50,295,726 bbl. Four wells that only produce oil generate small volumes of water that were not included in the total. Produced water volumes are not directly measured, but rather are estimated by oil producers.

Produced water is managed through underground injection in Florida. Twenty wells injected 15,533,899 bbl of produced water for disposal. Forty-three wells injected 34,761,872 bbl of water for enhanced recovery.

5.8. Illinois

The Illinois State Geological Survey (ISGS) provided information on average production between 2000 and 2006 (Dastgheib 2008). Reported data comes from the ISGS waterflood database. Annual oil production from enhanced oil recovery (EOR) averaged 3,201,958 bbl during that period. Produced water averaged 136,872,199 bbl annually resulting in a WOR of 42 7 bbl/bbl. As the database contains only information related to waterflooding, production activities in Illinois that do not use waterflooding would not be included. The actual produced water volume in the state is likely to be higher than what is reported here, as EIA reported oil production for Illinois was 9,609,000 bbl for 2007 (EIA 2008a).

Nearly all of the generated produced water from EOR is injected for waterflooding activities to improve recovery of crude oil. 135,263,569 bbl of the produced water was managed in this way. It is unclear from the available information whether produced water is also managed via injection for disposal. Additional produced water may be disposed of by methods other than waterflooding, although data on that is unknown.

5.9. Indiana

Produced water data were provided by the Division of Oil and Gas of the Indiana Department of Natural Resources (IDNR) and estimated from reported injections for 2007 (Nemecek 2008). Additional production information was obtained through the December 2007 monthly report published by the IDNR (IDNR 2007). Total reported production for 2007 was 1,726,553 bbl of crude oil and 3,605.982 Mmcf of gas. While generation information on produced water was not available, management information states that 40,200,000 bbl of produced water was managed through injection in 2007. The primary management practice for produced water was injection for enhanced recovery, which injected 34,500,000 bbl of produced water annually into 1,101 injection wells. Produced water is also managed through 183 disposal wells, which managed the remaining 5,700,000 bbl of produced water.

5.10. Kansas

Production data were provided by the Kansas Corporation Commission (KCC) and obtained from the Kansas Geological Survey's 2007 report (Durrant 2008). The 2007 drilled wells and CBM wells are from the 2007 KCC ACO-1 well completion forms. The produced water volumes are subjective and were determined using a weighted average approach from injection well permits in Kansas. The injection well permit information was collected from the U3-C annual reports of pressure monitoring, fluid injection, and enhanced recovery filed by operators.

In 2007, 67,631 wells produced hydrocarbons in Kansas. There were 43,384 wells producing 36,611,778 bbl of crude oil. Conventional gas production occurred at 24,247 wells and totaled 370,918.167 Mmcf. Additionally, 1,306 wells were drilled in 2007 to begin CBM production. Crude oil production generated 796,370,268 bbl of produced water, resulting in a WOR of 22 bbl/bbl. Conventional gas production generated 447,958,276 bbl of produced water, resulting in a WGR of 1,207 bbl/Mmcf. The total volume of produced water generated in Kansas was 1,244,328,544 bbl.

Produced water is managed via injection wells in Kansas. Offsite commercial disposal of produced water is not a recognized management practice in Kansas. There were 11,016 injection wells for enhanced recovery and 4,406 injection wells for disposal in 2007. Disposal is the primary purpose for injection, with 800,009,421 bbl of produced water managed in this way. Injection for enhanced recovery managed 444,319,123 bbl of produced water.

5.11. Kentucky

The Kentucky Division of Oil and Gas Resources supplied only hydrocarbon production information, as it does not have regulatory authority to monitor produced water. There were 49,344 wells in Kentucky in 2007. Crude oil was produced from 26,481 wells, which generated 3,572,400 bbl of crude oil (Combs 2008). Gas production occurred at 19,385 wells, which produced 95,246.846 Mmcf for gas, according to the Kentucky Geological Survey (undated). There were an additional 3,478 wells that include storage, stratification tests, and combination oil and gas wells. No information was provided on produced water generation or management.

Produced water volumes were estimated by extrapolation using available production information and water-to-hydrocarbon ratios from neighboring states. The WGR (17.7 bbl/Mmcf) from CBM activities in Virginia was multiplied by the gas production in West Virginia (225,000 Mmcf) to estimate the volume of water attributable to West Virginia's gas production (3,982,500). This gas-specific water volume was subtracted from the total West Virginia water volume (8,337,000 bbl) to estimate the volume of water attributed to West Virginia crude oil production (4,354,500 bbl). This oil-specific water volume was divided by the West Virginia crude oil production to estimate the WOR in West Virginia (6.4 bbl/bbl). The WGR from Virginia and the estimated WOR in West Virginia were used to determine the total estimated produced water volume for Kentucky. Based upon the previous assumptions, produced water generated from oil production would be 22,920,000 bbl. Produced water generated from gas production in Kentucky is estimated to be 1,688,000 bbl. Total produced water generation for 2007 is estimated at 24,608,000 bbl.

5.12. Louisiana

Hydrocarbon production and produced water management information was provided by the Louisiana Department of Natural Resources (Parker 2008). Louisiana does not track the amount of water produced from each of Louisiana's 80,892 actively producing wells. Of these wells, 40,957 produce crude oil and 39,935 produce gas. Total oil production in 2007 was 52,495,100 bbl. Gas production volume was 1,381,585 Mmcf.

While production volume of produced water is unavailable, Louisiana does track injection volumes for 2,914 wells (1,149,643,443 bbl of produced water in 2007). Ninety percent of produced water (1,034,092,270 bbl) was injected for disposal. The remaining water was either managed by injection for enhanced recovery (66,261,179 bbl of water into 354 wells) or by offsite commercial disposal (49,289,994 bbl of water into 31 wells).

5.13. Michigan

The Michigan Department of Environmental Quality's Office of Geological Survey provided information on produced water management (Organek 2008). Production volume data was obtained from a database on the Gas, Oil, and Minerals website of the Michigan Department of Environmental Quality. Total crude oil production was 5,180,043 bbl. Michigan also produced 441,844 bbl of condensate and 453,749 bbl of natural gas liquids in 2007. Gas production during that period was 166,794.419 Mmcf. Produced water generation for all hydrocarbon production activities was 114,580,484 bbl.

Michigan manages produced water through underground injection. Twenty percent of produced water was injected for enhanced recovery, while the majority of produced water (80%) was managed through injection for disposal.

5.14. Mississippi

The Mississippi Oil and Gas Board provided produced water generation and management information (Sims and Tarbutton 2008). Mississippi has three categories of production wells: wells that produce only oil, wells that produce only gas, and wells that produce both oil and gas. For oil-only wells, 1,379 wells produced 12,111,145 bbl per year of crude oil and 164,027,705 bbl of produced water. The WOR for these wells was 13.54 bbl/bbl. There are 1,403 gas-only wells in Mississippi, which produced 40,656 Mmcf of gas and 1,459,563 bbl of produced water, with a WGR of 35.9 bbl/Mmcf. The remaining 1,903 wells produce both oil and gas, producing 7,915,632 bbl of crude oil, 56,056 Mmcf of gas, and 165,242,314 bbl of water in 2007. Total oil production was 20,026,777 bbl, and total gas production was 96,712 Mmcf. Total produced water generation in Mississippi was 330,739,582 bbl for 2007.

Total produced water managed in Mississippi exceeds the total produced water generated, which suggests that most of the water injected for enhanced recovery consists of makeup water. There were 381 wells that injected 389,613,955 bbl of water for enhanced recovery. In addition, 466 wells injected 281,563,270 bbl of water for disposal.

5.15. Missouri

The Division of Geology and Land Survey of the Missouri Department of Natural Resources provided produced water generation and management information (Kaden 2008). In 2007, Missouri had 308 crude oil production wells. There was no commercial gas production activity in Missouri for that year. The 308 wells produced 79,516 bbl of oil and 1,612,592 bbl of water, for a WOR of 20.28 bbl/bbl. Injection for disposal is the primary management method for produced water. One hundred forty-six injection wells managed 1,611,246 bbl of produced water. The remaining 1,346 bbl of water were managed through evaporation.

5.16. Montana

The Montana Board of Oil and Gas provided produced water generation and management information (Halvorson 2009; Hudak 2008). In 2007, crude oil production volumes were 34,749,250 bbl from 4,325 wells. Gas production took place at 5,694 wells and yielded 82,119 Mmcf. Much of the current and historic gas production in Montana occurred in formations that would be considered unconventional, as tight sand or siltstone reservoirs, low-volume gas wells, or "shale gas," in some cases. As a result, Montana does not break out gas sources as conventional or unconventional. CBM production information is collected separately, and 902 wells produced 13,062 Mmcf of CBM. Produced water volumes from these hydrocarbons were the following: 139,189,532 bbl of water from crude oil production, 4,196,725 bbl of water from gas production, and 38,880,091 bbl from CBM production. These volumes resulted in a WOR of 4.0 bbl/bbl, a WGR of 51.1 bbl/Mmcf of gas, and a WGR of 2,976 bbl/Mmcf of CBM. When conventional natural gas and CBM are combined, the resulting WGR is 453 bbl/Mmcf.

The majority of produced water generated is managed through injection wells. There were 624 active wells for enhanced recovery and 173 wells for disposal. Enhanced recovery managed 60% (109,217,415 bbl) of the produced water generated in Montana. Twenty-six percent (46,807,463 bbl) of produced water generated was managed through injection for disposal. The remaining 14% of produced water was managed in other ways including the following: evaporation pits, beneficial use including stock water and dust suppression, and direct discharge to surface waters. According to Keith et al. (2003), produced water from CBM activities was primarily managed through other practices including direct discharge, holding ponds, and land application.

5.17. Nebraska

The Nebraska Oil and Gas Conservation Commission provided both production and management information (Sydow 2008). In 2007, there were 1,384 wells, with 1,187 of those wells producing oil. The remaining 197 wells produced gas. Oil production was 2,335,355 bbl and generated 48,835,874 bbl of produced water (WOR: 20.91bbl/bbl). Conventional gas production volume was 1,331.125 Mmcf and generated 476,481 bbl of produced water (WGR: 357.95 bbl/Mmcf). Total produced water generation in Nebraska was 49,312,355 bbl.

For produced water management, data were provided both according to the number of wells that produce water and the number of injection wells that receive the produced water. The majority of produced water (31,588,486 bbl) was sent from 673 oil wells to be managed by 407 injection wells for enhanced recovery. Injection for disposal was conducted at 91 injection wells to manage 14,337,458 bbl from 478 oil wells and 197 gas wells. Sixteen oil wells sent 499,960 bbl of produced water to surface discharge. Evaporation management took 2,886,451 bbl of produced water from 233 oil wells.

5.18. Nevada

The Nevada Division of Minerals supplied production data (Price 2009). According to the 2007 bimonthly reports of the *Nevada Oil Patch*, total oil production in 2007 was 408,174 bbl. Gas production for the same year was 4.474 Mmcf. Produced water volumes were collected for oil production, and 6,784,661 bbl were generated in 2007. No information on management of produced water was provided, although a well inventory in 2004 revealed that there were 13 Class II injection wells in Nevada (NDEP 2004).

5.19. New Mexico

The Oil Conservation Division of the New Mexico Energy, Minerals, and Natural Resources Department provided produced water volume information according to hydrocarbon production (Fesmire 2009). Crude oil production generated 53,265,771 bbl of oil and 534,273,532 bbl of water in 2007 (WOR: 10.0 bbl/bbl). An additional 5,872,258 bbl of oil condensate were produced; total oil production for 2007 was 59,138,029 bbl. Conventional gas production totaled 797,747 Mmcf and generated 80,377,476 bbl of produced water (WGR: 100 bbl/Mmcf). Coal bed methane production was 496,825 Mmcf and produced 38,084,472 bbl of water (WGR: 76.7 bbl/Mmcf). When conventional natural gas and CBM are combined, the resulting WGR is 92 bbl/Mmcf.

While unconventional gas and condensate were produced (231,901 Mmcf and 5,872,258 bbl, respectively), the volume of produced water from these activities was unknown. Total gas production in 2007 was 1,526,473 Mmcf. Additional produced water was generated from carbon dioxide production activities. For 98,856,249 Mmcf of carbon dioxide, 65,585 bbl of produced water were generated. The total produced water from the above activities was 652,801,065 bbl, although the total water actually produced in the state in 2007 was 665,684,732 bbl. The remaining amounts could not be identified by these hydrocarbon categories.

New Mexico does not specifically track the produced water that gets reinjected. While the option is included in the monthly production reports, well operators use different categories for the same action. In 2007, 3,167 wells had 449,488,784 bbl of water injected specifically into injection wells. An additional 603 wells had 348,141,541 bbl of water injected as salt water disposal.

5.20. New York

The New York State Department of Environmental Conservation's Division of Mineral Resources provides production information in an annual report and through an online database (NYSDEC 2006; NYSDEC undated). The most recent available report is for 2007. According to the 2007 data, 13,113 wells were reported to the division. Of the total, 7,387 were natural gas wells, 4,874 were oil wells, and the remaining wells were gas storage, dry holes, and solution salt wells. The database provided production volumes of 55,001 Mmcf for natural gas and 377,514 bbl for oil. The state-produced water volume was 649,333 bbl from active wells for 2007, which included 215,050 bbl that were associated with water injection wells.

Specific details were not available on the management practices for produced water in New York, although injection wells for disposal and enhanced recovery were listed under well type codes within the database. All produced water that is hauled from well sites requires a department-issued waste transporter permit. The department allows produced water disposal at publicly owned treatment works, with approved industrial pretreatment or mini-pretreatment programs. The department also allows road spreading of production brine for deicing and dust control purposes under its beneficial use determination program.

5.21. North Dakota

The Department of Mineral Resources provides oil and gas production information online. In 2007, there were 3,872 oil producing wells in North Dakota. Of these wells, 2,525 also reported gas production. In addition to the 3,872 oil wells, there were 216 gas wells. Fifty-nine of the gas wells were permitted as gas condensate wells and produced some oil. Total annual crude oil production from those wells was 44,701,934 bbl. The oil producing wells, which generated 44,543,187 bbl of the total crude oil, generated 134,704,290 bbl of produced water (WOR: 3.02 bbl/bbl). Gas production for 2007 was 70,767.153 Mmcf. Much of this production volume was coproduced from oil wells. The gas wells produced 15,909 Mmcf in 2007 and generated 286,889 bbl of produced water (WGR: 18.03 bbl/Mmcf). The total produced water generated in North Dakota from these production activities was 134,991,179 bbl.

More than 96% of the produced water generated in North Dakota is managed through injection. In 2007, there was slightly more produced water managed through injection for disposal than enhanced recovery. Water produced from 2,531 wells totaling 65,321,067 bbl was directed to disposal wells, which managed 48.39% of the total volume of produced water for 2007. Enhanced recovery managed 64,872,788 bbl (48.06% of the total volume) of produced water from 1,096 wells. The remaining 3.55% of the produced water (4,797,324 bbl) in North Dakota came from 519 wells and was directed to offsite commercial disposal wells.

5.22. Ohio

The Ohio Division of Mineral Resources Management provided produced water information for 2006, and production data was obtained for the same year through the 2006 report *Summary of Ohio Oil and Gas Activities* (ODNR 2006). The majority of Ohio wells are combination wells, producing both oil and gas. In 2006, there were 63,654 active wells in Ohio, which produced 5,422,194 bbl of crude oil and 86,315.100 Mmcf of natural gas. These wells produced 6,940,000 bbl of produced water in 2006. Ohio does not subcategorize reported produced water volumes by hydrocarbon.

Most produced water generated in Ohio is managed via injection for disposal. Eighty-eight percent, 6,137,005 bbl, of reported produced water in 2007 was injected for disposal. Injection for enhanced recovery managed 7% of the produced water generated (487,478 bbl). The remaining 315,517 bbl were managed either at annular disposal wells or were reused through spreading for dust and ice control with local government authorization.

5.23. Oklahoma

The Oklahoma Corporation Commission (OCC) publishes an annual report on oil and gas. Production information for 2007 was obtained from the 2008 report, and produced water management information was provided directly by the OCC. For the reported year, there were 82,832 crude oil production wells producing 60,760,000 bbl of crude oil. Gas production occurred at 38,364 wells and totaled 1,643,292 Mmcf.

The majority of produced water in Oklahoma was managed via injection in 2007 (Baker 2009). Injection occurred at 2,584 active disposal wells, which managed 1,201,680,455 bbl of produced water. Injection also occurred at 202 commercial disposal wells that managed 52,451,784 bbl of produced water. In total, injection for disposal wells managed 1,254,132,239 bbl. Oklahoma had 6,264 enhanced recovery wells in 2007, which managed 940,271,840 bbl of produced water. Currently, one ongoing surface discharge operation in the Arkoma Basin discharged 165 bbl in 2008 (note: 2007 data not available). Oklahoma also has one recycling permit issued for produced water. The permit is for a surface reclamation project in which produced water is mixed with fresh water. The mixed water is combined with fly ash to be used as fill for closure at abandoned strip mines. The amount of water reported as used in this operation for 2007 was 776,200 bbl. The total produced water managed in Oklahoma for 2007 (including the 2008 surface discharge volume as an estimate) was 2,195,180,444 bbl.

5.24. Pennsylvania

Production information is available via the Wells Information System (WIS) created by the Pennsylvania Geological Survey and provided by the Pennsylvania Department of Environmental Protection's Bureau of Oil and Gas Management (PDEP 2008a). In 2007, total gas production was

172,367.077 Mmcf, while reported oil production was 1,537,347 bbl. Production occurred at 65,129 wells. Estimates of production wells were slightly higher but broken into categories by hydrocarbon (PDEP 2008b). Gas production wells totaled 52,700 wells, and there were 18,200 oil wells in 2007. The total produced water volume from these production wells was 3,912,456 bbl. Information on produced water management was not provided.

5.25. South Dakota

The Department of Natural Resources Minerals and Mining Program (Oil and Gas Section) provided information on production and management of produced water related to oil and gas activities (McGillivray 2008). In 2007, a total of 224 wells were active in South Dakota, with 153 producing crude oil and 71 producing conventional gas. Oil production totaled 1,664,889 bbl and generated most of the produced water generated in South Dakota, with 4,185,565 bbl attributable to oil production (WOR: 2.5 bbl/bbl). Gas production generated only 486 bbl of produced water in generating 11,880 Mmcf of conventional gas (WGR: 0.04 bbl/Mmcf).

Most produced water is managed through injection. Ninety-six percent of produced water generated in South Dakota was injected, with 2,121,947 bbl injected for enhanced recovery (51%) and 1,853,200 bbl injected for disposal (45%). The remaining four percent of produced water was managed either through surface discharge (85,243 bbl) or beneficial reuse as water for livestock (85,243 bbl).

5.26. Tennessee

Tennessee does not presently keep track of produced water information from oil and gas activities. According to Tennessee's Division of Geology's website, approximately 350,000 bbl per year of oil and 1,000 Mmcf per year of gas are produced in the state (TDEC undated). Produced water management information was unavailable. However, according to U.S. EPA Region 4, as of August of 2007, there were 13 Class II injection wells in Tennessee (U.S. EPA 2008b).

Produced water volumes were estimated by extrapolation using available production information and water-to-hydrocarbon ratios from neighboring states. The water-to-gas ratio (17.7 bbl/Mmcf) from CBM activities in Virginia was used to determine the WOR in West Virginia (6.4 bbl/bbl) (see discussion for Kentucky, above). These ratios were assumed to be the same for Tennessee, with the exception of CBM, as no CBM activities were explicitly stated in the production information obtained from the state. Based upon the previous assumptions, produced water generated from oil production would be 2,246,000 bbl. Produced water generated from gas production in Tennessee is estimated to be 18,000 bbl. Total produced water generation for 2007 is estimated at 2,263,000 bbl.

5.27. Texas

Produced water injection data for 2007 were provided by the Railroad Commission of Texas, and hydrocarbon production data was obtained from the commission's website. In 2007, Texas wells produced 342,086,945 bbl of crude oil and 6,878,016 Mmcf of natural gas (RRC 2009). There were more than 216,000 active oil and gas wells statewide, which managed produced water via injection into more than 50,000 permitted oil and gas injection and disposal wells (RRC undated). While CBM production activities occur in Texas, such activities are not distinguished from conventional gas production (Deleon 2009).

Currently, injection reports for 2007 are still being added to the commission's database. As of April 2, 2009, the total reported liquid injection volume for 2007 was 4,254,037,585 bbl. From the current total, 1,364,096,989 bbl (32% of reported total) were disposed into nonproducing formations, 757,606,954 bbl were (18% of reported total) disposed into producing formations, and 2,132,117,391 bbl (50% of reported total) were injected for enhanced recovery. For the sake of comparison to other state totals, we combined the percentage disposed into producing formations with the percentage injected for enhanced recovery (68%). The projected volume for the state for 2007 is 7,376,912,883 bbl (Deleon 2009). The total volume injected includes salt water, brackish water, and fresh water injection into Class II disposal wells. While a small amount of produced water was managed through tidal discharge, the vast majority of produced water generated in Texas was managed through injection. We assume the projected injected volume for 2007 (7,376,912,883 bbl) is comparable to the actual volume of produced water generated in Texas. Following the percentage allocation assumption described above, we considered 32% of the projected total (2,360,612,122 bbl) as the volume injected for disposal. We considered 68%

of the projected total (5,016,300,761 bbl) as the volume for enhanced recovery.

5.28. Utah

The Utah Division of Oil, Gas, and Mining provided estimates according to data available as of December 2007 (Kierst 2008). 8,788 wells were operating in the state in 2007, with 3,022 wells producing crude oil. It was estimated that 4,401 wells were producing conventional gas, with the remaining 1,035 wells producing CBM. Crude oil production in 2007 was 19,519,964 bbl. Conventional gas production during the same period was 311,741 Mmcf, and 73,638 Mmcf of CBM was produced. While produced water volumes are not available according to hydrocarbon type, annual produced water generation for 2007 was 148,579,384 bbl.

The majority of produced water is managed through injection (Kierst 2009). There were 1,085 wells used for injection for enhanced recovery; the produced water (estimated to be 78,251,086 bbl) was comingled with makeup water. Ninety-three wells injected produced water for disposal; in 2007, these wells injected 60,157,307 bbl. Commercial disposal wells managed an additional 1,893,393 bbl of produced water, assuming no overlap in reporting. Utah also has 15 surface discharges in the Ashley Valley Field. The produced water generated in this field (21,079,950 bbl in 2007) serves local agriculture and eventually reaches the Green River. It is likely that much of the volume reported for evaporation ponds (7,773,128 bbl) overlapped the reporting by commercial evaporation ponds, which were estimated to manage 11,067,743 bbl. Drilling and workover operations managed 476,945 bbl of produced water. Onsite evaporative pits on federal lands were estimated to manage 218,239 bbl of produced water.

5.29. Virginia

Virginia requires produced water data to be reported for CBM operations only, and the Virginia Department of Mines, Minerals, and Energy (VA DMME) provided production information concerning CBM activities (Kent 2008). In 2007, 4,290 wells were producing 88,146.284 Mmcf of CBM and 1,561,842 bbl of produced water (WGR: 17.7 bbl/Mmcf). Production information for oil, conventional gas, and coproduced gas for 2007 was obtained from available county production statistics (VA DMME undated). Total gas production in 2007 was 112,000 Mmcf, with 79% of that volume attributed to CBM activities. Conventional gas activities produced approximately 23,505.106 Mmcf of gas (VA DMME undated). Oil production volume in 2007 was 19,155 bbl (VA DMME undated). Virginia is beginning to collect data on produced water management strategies including injection for disposal and surface discharge of produced pit fluids, which should be available electronically in the future.

5.30. West Virginia

The West Virginia Office of Oil and Gas provided information on production activities and produced water management (Martin 2008). In 2007, there were 49,618 producing wells in West Virginia, with the majority of those wells (45,398) producing unconventional gas. The remaining wells produced crude oil (3,631) and CBM (589). Unconventional gas production for 2007 was 201,599 Mmcf, and CBM production was 22,974 Mmcf. During the same period, 679,239 bbl of crude oil were produced. Total produced water generation from these activities is not available.

While generation information is not available, produced water management information can provide an estimate for generation within the state. There were 709 injection wells that managed 3,942,370 bbl of produced water for enhanced recovery. An additional 69 injection wells disposed of produced water, although the total volume managed is not available. West Virginia managed 3,857,000 bbl of produced water from CBM activities through 100 permitted land applications, which allowed produced water of a certain quality to be dispersed on the ground. An additional 537,659 bbl of produced water was managed through offsite commercial disposal. While the total volume for disposal wells is unknown, the known volume of produced water in West Virginia in 2007 was at least 8,337,029 bbl.

5.31. Wyoming

The Wyoming Oil and Gas Conservation Commission provided historic information on produced water management from the Powder River Basin and qualitative information on reuse of produced water in the Big Horn Basin (Likwartz 2008). Specific information on production was obtained from statistics available on the commission's website (WOGCC undated). In 2007, 54,051,671 bbl of crude oil were produced. Total gas production reached 2,253,487.217 Mmcf, with 436,306.044 Mmcf of CBM. Production of oil and gas resulted in the generation of 2,355,671,186 bbl of produced water in 2007.

Information on historic trends that was provided for produced water management in the Powder River Basin indicate that since 1987, 4.784 billion bbl have been produced from coal beds. Approximately 54% has been discharged to ephemeral and perennial streams, 35% has been managed using off-channel pits, 5% has been reused for irrigation projects, 3% has been managed through injection, and 3% has been treated and then discharged into streams. Much of the produced water from conventional gas activities in the Big Horn Basin was also managed through agricultural reuse, although the actual volume is unknown.

6. Federal and Tribal Summary

This section provides detailed information on produced water associated with production activities on federal lands (onshore), federal offshore, and tribal lands. Federal onshore mineral leasing activities are managed by the U.S. Department of the Interior's Bureau of Land Management and the U.S. Department of Agriculture's Forest Service. The MMS manages the mineral resources (e.g., oil and gas) on the outer continental shelf. The Minerals Revenue Management Program (MRM), which is within MMS, is responsible for management of all revenues associated with mineral leases on federal onshore, federal offshore, and tribal lands. The information in this chapter was obtained from MMS.

6.1. Federal Onshore Production

Onshore production activities on federal lands are assumed to be included in the total production volumes provided by the state agencies. In 2007, oil and gas production on federal lands were 104,378,709 bbl and 2,889,060 Mmcf, respectively. These production volumes are included in the state summaries in the previous chapter. The volumes represented 8.2% of state oil production and 13.5% of state gas production. Produced water volumes from federal onshore production activities are part of the more than 20 billion barrels of produced water from state production totals and were not separately reported.

6.2. Federal Offshore Production

In the produced water generation and management information for federal offshore activities the MMS provided (Kazanis et al. 2009), the number of wells provided was actually the number of completion counts, as some wells had both oil (3,268) and gas completions (3,580). Oil produced from oil completions totaled 385,066,136 bbl and generated 399,820,418 bbl of produced water (WOR: 1.04 bbl/bbl). The volume of gas produced from gas completions was 2,179,348.290 Mmcf, which generated 187,532,428 bbl of produced water (WGR: 86.0 bbl/Mmcf). Total oil production from both oil and gas completions was 467,179,615 bbl. Total gas production in 2007 was 2,786,592.780 Mmcf. Produced water generated from all production activities was 587,352,846 bbl.

While some produced water generated from offshore activities was injected, the vast majority of produced water was managed through discharge. In 2007, injection for enhanced recovery managed 48,673,102 bbl and injection for disposal managed 1,298,417 bbl of produced water. The remaining produced water (537,381,327 bbl) was treated and then discharged.

6.3. Tribal Lands

Forty-two tribes are served by MRM, and MRM provides total annual production and revenue data for the tribes online. The production data and national average water-to-hydrocarbon ratios were used to estimate produced water generation on tribal lands. The following tribes are those that are included in the annual production summary:

- Alabama-Coushatta
- Arapahoe
- Assiniboine Sioux
- Blackfeet
- Caddo
- Cherokee
- Cheyenne-Arapaho
- Chickasaw
- Chilocco Indian School
- Chippewa-Cree
- Navajo
- Pala Band Mission Indians
- Pawnee
- Ponca
- Pueblo of San Ildefonso
- Pueblo of Sandia
- Pueblo of Zia
- Pyramid Lake Paiute
- Quechan
- Sac and Fox

- Choctaw
- Colorado River
- Cook Inlet Region, Inc.
- Creek
- Crow
- Delaware
- Ft. Mohave
- Gila River Indian Community
- Hopi
- Jicarilla Apache
- Morongo Band Mission Indians
- Salt River-Pima Maricopa
- Santa Ana Pueblo
- Seminole
- Shoshone
- Shoshone Bannock
- Soboba Band Mission Indians
- Southern Ute
- Tohono O'Odham
- Ute
- Ute Mountain Ute
- Wichita

MRM reports that in 2007 oil production was 9,512,652 bbl and gas production was 297,329 Mmcf. Location-specific production information was not obtained; therefore, the U.S. averages for WOR (7.57) and WGR (259.8) were used to estimate the produced water volume for tribal lands. Total produced water was estimated to be 145,109,593 bbl.

7. FINDINGS AND CONCLUSIONS

7.1. Findings

7.1.1. Produced Water Volume

This report provides a current estimate for the volume of produced water generated from oil and gas production in the United States. The volume estimate represents a compilation of data obtained from numerous state oil and gas agencies and several federal agencies for 2007, where possible. The total volume of produced water estimated for 2007 is about 21 billion bbl. This equals an average of 57.4 million bbl/day. Produced water is generated from most of the nearly 1 million actively producing oil and gas wells in the United States.

Not all states had readily available precise produced water volume figures. In a few states, the agencies had very complete data records easily obtainable from online sources. Other states had summary-level volume data without much detail or had data available only in in-house data repositories. The most challenging states were those that had no produced water data at all. In those cases, we calculated estimates through extrapolation and correlations using hydrocarbon production and produced water volumes from neighboring states.

Several states dominate the total produced water volume estimates. Texas, with more than 7.3 billion bbl, represented 35% of the national total. Other states with produced water volumes exceeding 1 billion bbl include California (12%), Wyoming (11%), Oklahoma (11%), Kansas (6%), and Louisiana (5%).

The greatest produced water contributors are not necessarily the greatest producers of oil and gas. While Texas was the largest gas producer in the United States (nearly 6,900,000 Mmcf in 2007), federal offshore production activities provided the largest volume of crude, more than 467,000,000 bbl. Although federal offshore production generated nearly 27% of U.S. crude oil production, less than 3% of total U.S. produced water was generated from federal offshore activities.

7.1.2. Produced Water Volume by Hydrocarbon Type

We asked the agencies to provide produced water volumes by hydrocarbon types (i.e., crude oil, conventional gas, coal bed methane, unconventional gas, or other), to the extent the data were available at that level of detail. Most states were unable to break out produced water volumes for all categories, but some states could at least provide estimates of produced water from oil production vs. natural gas production. States that segregated produced water by hydrocarbon categorized 6,666,144,270 bbl of produced water. Eighty-seven percent (5,770,327,439 bbl) of the produced water came from oil production activities.

It is useful to consider the WOR and WGR from production activities to evaluate the relative production age of resources within the production lifetime. We were able to calculate separate WORs and WGRs for several states that reported produced water separately by hydrocarbon type (WORs for 14 states and WGRs for 11 states).

For WORs, state values ranged from 2.5 bbl/bbl for South Dakota to 43 bbl/bbl for Illinois. The offshore ratio was even smaller at 1.04 bbl/bbl. A national average ratio, calculated using the production-weighted ratios from the 14 states, was 7.6 bbl/bbl. When offshore production was added to the onshore production, the total average U.S. WOR was 5.3 bbl/bbl.

For WGR ratios, the state values ranged from 0.04 bbl/Mmcf in South Dakota to more than 1,200 bbl/Mmcf for Kansas. The offshore ratio was 86 bbl/Mmcf. A national average ratio calculated using the production-weighted ratios from the 11 states was 260 bbl/Mmcf. When offshore production was added to the onshore production, the total average U.S. WGR was 182 bbl/Mmcf.

7.1.3. Produced Water Management Practices

This report describes the current (i.e., 2007) practices used by many oil and gas producers to manage produced water. Two general themes are followed by most U.S. operators. More than 98% of produced water from onshore wells is injected underground. Approximately 59% is injected into producing formations to maintain formation pressure and increase the output of production wells. Another 40% of produced water from onshore wells is injected into nonproducing formations for disposal.

Four percent of the total reported volume of produced water (onshore and offshore) managed in 2007 was managed through surface discharges. Alabama is noteworthy among states with onshore production because it managed about 65% of its produced water through surface discharge. Nearly all of the discharges came from CBM operations in the Black Warrior Basin. The remaining produced water volume was managed through evaporation ponds, offsite commercial disposal, beneficial reuse, and other management methods.

More than 91% of offshore produced water, including the water from inshore platforms in Cook Inlet, Alaska, was discharged to the ocean. Most of the remaining volume was reinjected for enhanced recovery.

One of the potentially valuable management practices for produced water is beneficial reuse. We attempted to obtain information on the volume of produced water reused and how the water was reused, but were unsuccessful at finding much information.

We had hoped to obtain produced water management information separately for each type of hydrocarbon production. Most states were either not able to provide that information or reported that the same practices were used regardless of the hydrocarbon type.

7.2. Conclusions

This report provides the most detailed and current information on the volume of produced water generated in the United States. Its 2007 estimate was derived in a more detailed and comprehensive manner than earlier national estimates developed for 1985, 1995, and 2002. The 2007 volume is the highest of those estimates, and should be a more representative estimate than the earlier ones because of the approach used to collect and extrapolate the data.

Information on management practices has not changed significantly from previous studies. However, confirmation of that fact is valuable, too. The majority of produced water is managed through injection with 59% of the onshore produced water volume being injected to increase production output.

A final important conclusion of this study is that there is no easy way to obtain national estimates of produced water generation. The estimates presented in this report took months of investigation, numerous contacts with oil and gas agency staff members, and extensive follow-up. Some states had useful produced water information either published in reports or readily available through state databases. However, many other states had only minimal information about produced water volumes or how the produced water was managed. No federal regulatory program requires agencies to track produced water volume. Consequently, when regulatory and data management resources are limited, some states do not maintain produced water information.

Appendix A. Sample Letter

November 7, 2008

[Title]. [First Name] [Last Name]
State Oil and Gas Agency
123 State Street
City, State 12345

Dear [Title]. [Last Name]:

The U.S. Department of Energy (DOE) asked Argonne to compile data on two important water streams associated with oil and gas production – produced water and hydraulic fracturing (frac flowback) water. No recent national volume estimates of these water streams are available, nor do we have satisfactory information on how the water is managed after it is generated. Argonne plans to develop detailed national-level information on the volume of produced water generated in the United States and the ways in which it is disposed or reused. In a second study, Argonne will evaluate frac flowback water use and re-use to determine the sources of water used for fracturing and the methods used to manage the flowback water after it is returned to the surface.

We plan to collect data from various sources, but believe that the State oil and gas programs are the best place to begin. Your agency has long-term direct experience with oil and gas activities in your state. Further, most states already employ data management systems for tracking production data.

I know that you are very busy with your regular responsibilities, but I hope that you can designate a member of your staff to work with us to identify the types of relevant information that we can find from your existing files and databases. If necessary, we can send one of our project team members to your offices to review and compile data. We would like to begin these two projects during November. Please let me know whom we can contact in your agency to obtain the necessary data.

Although I am overseeing the projects, much of the work will be done by my colleague, Dr. Corrie Clark. You may contact either me (xxx-xxx-xxxx; xxx@anl.gov) or Dr. Clark (xxx-xxx-xxxx; xxx@anl.gov). Thank you in advance. We look forward to working with you.

Sincerely,

John A. Veil

References

ALL (ALL Consulting), (2003). *Handbook on Coal Bed Methane Produced Water: Management and Beneficial Use Alternatives*, prepared by ALL Consulting for the Ground Water Protection Research Foundation, U.S. Department of Energy, and U.S. Bureau of Land Management, July 2003.

Alaska OGC (Alaska Oil and Gas Commission), (2009). Archived Monthly Production Reports for 2007. Available at http://www.aogcc.alaska.gov/production/ProdArchives/parchiveindex.shtml. Accessed April 30, 2009.

API (American Petroleum Institute), (1988). "*Production Waste Survey*," prepared by Paul G. Wakim, June.

API, (2000). *Overview of Exploration and Production Waste Volumes and Waste Management Practices in the United States*, prepared by ICF Consulting for the American Petroleum Institute, Washington, DC, May.

Arkansas OGC (Arkansas Oil and Gas Commission), (2007). *Annual Report of Production: 2007*. Available at http://www.aogc.state.ar.us/ OnlineData /reports/Annual%20Report%20of%20Production%201907%20-%202007.pdf. Accessed April 6, 2009.

Baker, T., (2009). personal communication among T. Baker, and L. Wrotenbery, Oklahoma Corporation Commission, Oklahoma City, OK, and J. Veil and C. Clark, Argonne National Laboratory, Washington, DC, May 14.

Benko, K., & Drewes, J. (2008). "Produced Water in the Western United States: Geographical Distribution, Occurrence, and Composition," *Environmental Engineering Science 25*, (2), 239–246.

Bolin, D., (2008). personal communication between D. Bolin, State Oil and Gas Board of Alabama, Tuscaloosa, AL, and C. Clark, Argonne National Laboratory, Washington, DC, Dec. 23.

Breit, G., Klett, T. R., Rice, C. A., Ferderer, D. A. & Kharaka, Y. (1998). "*National Compilation of Information about Water Co-produced with Oil and Gas*," presented at the 5th International Petroleum Environmental Conference, Albuquerque, NM, Oct. 20–23.

Brost, D. (2002). "*Water Quality Monitoring at the Kern River Field*," presented at the 2002 Ground Water Protection Council Produced Water Conference, Colorado Springs, CO, October 16–17. Available at http://www.gwpc.org/meetings/special/PW%202002/ Papers/Dale_Brost_PWC2002.pdf. Accessed May 5, 2009.

CDOC (California Department of Conservation), (2008a). *2007 Preliminary Report of California Oil and Gas Production Statistics*, Publication No. PR03, revised April 2008, Division of Oil, Gas, and Geothermal Resources. Available at ftp://ftp.consrv.ca.gov/pub/oil/ annual_reports /2007/PR03_2007.pdf. Accessed May 5, 2009.

CDOC, (2008b). "Online Production and Injection Database," Division of Oil, Gas, and Geothermal Resources. Available at http://opi.consrv.ca. gov/opi/opi.dll. Accessed April 6, 2009.

Chapelle, F. H., (2001), "Chapter 2: Microorganisms Present in the Ground-Water Environment," *Ground-Water Microbiology and Geochemistry, 2nd Edition*, John Wiley & Sons, Inc., New York, NY.

Cline, J. T., (1998). "*Treatment and Discharge of Produced Water for Deep Offshore Disposal*," presented at the API Produced Water Management Technical Forum and Exhibition, Lafayette, LA, Nov. 17–18.

COGCC (Colorado Oil and Gas Conservation Commission), (2009). "*COGIS – Production Data Inquiry*." Available at http://cogcc.state.co. us/. Accessed March 4, 2009.

Combs, M., (2008). personal communication between M. Combs, Kentucky Department of Natural Resource, Division of Oil and Gas Conservation, Frankfort, KY, and C. Clark, Argonne National Laboratory, Washington, DC, Dec. 18.

Dastgheib, S., (2008). personal communication between S. Dastgheib, Illinois State Geological Survey, University of Illinois at Urbana-Champaign, Champaign, IL, and J. Veil, Argonne National Laboratory, Washington, DC, Dec. 2.

DeFilippo, M. N., (2004). *Use of Produced Water in Recirculating Cooling Systems at Power Generating Facilities*, semiannual technical progress report prepared by the Electric Power Research Institute for the U.S. Department of Energy. Available at http://www.netl.doe.gov/ technologies/coalpower/ewr/water/pp-mgmt/pubs/41906/41906Cost BenefitAnalysis.pdf. Accessed May 5, 2009.

Deleon, F., (2009). personal communication among F. Deleon, Railroad Commission of Texas, Austin, TX, and J. Veil and C. Clark, Argonne National Laboratory, Washington, DC, April 17.

Dowd, S. E., Herman, D. C. & Maier, R. M. (2000). "Chapter 6: Aquatic and Extreme Environments" in *Environmental Microbiology*, Academic Press, San Diego, CA.

Durrant, S., (2008), personal communication between S. Durrant, Kansas Corporation Commission, Topeka, KS, and C. Clark, Argonne National Laboratory, Washington, DC, Dec. 15.

EIA (Energy Information Administration), (2008a). "*Crude Oil Production*." Available at http://tonto.eia.doe.gov/dnav/pet/pet_crd_ crpdn_adc_mbbl _a.htm. Accessed March 16, 2009.

EIA, (2008b). "*Natural Gas Gross Withdrawals and Production*." Available at http://tonto.eia.doe.gov/dnav/ng/ng_prod_sum_a_epg0_ fgw_mmcf_a.htm. Accessed March 16, 2009.

Fesmire, M., (2009). personal communication between M. Fesmire, New Mexico Energy, Minerals, and Natural Resources Department, Oil Conservation Division, Santa Fe, NM, and C. Clark, Argonne National Laboratory, Washington, DC, Jan. 19.

Halvorson, J., (2009), personal communication between J. Halvorson, Montana Board of Oil and Gas, Billings, MT, and C. Clark, Argonne National Laboratory, Washington, DC, Jan. 14.

Hartig, L., (2009). letter from L. Hartig, Alaska Department of Environmental Conservation, Juneau, AK, to G. Stevens and M. Chenault, Alaska State Legislature, Juneau, AK, January 8. Available at http://www.dec. state.ak.us/water/npdes/pdfs/ReportLeg2009.pdf. Accessed April 15, 2009.

Hudak, G., (2008). personal communication between G. Hudak, Montana Board of Oil and Gas, Billings, MT, and C. Clark, Argonne National Laboratory, Washington, DC, Dec. 2.

Hutson, S. S., Barber, N. L. Kenny, J. F. Linsey, K. S. Lumia, D. S. & Maupin, M. A. (2004). "Estimated Use of Water in the United States in 2000," U.S. Geological Survey Circular 1268, 46 pp.

IDNR (Indiana Department of Natural Resources), (2007). *Monthly Report 12/31/2007*, Division of Oil and Gas. Available at http://www.in. gov/dnr/dnroil/files/mon0712.pdf. Accessed April 6, 2009.

Kaden, S., (2008). personal communication among S. Kaden and J. Jaquess, Missouri Department of Natural Resources, Division of Geology and Land Survey, Rolla, MO, and C. Clark, Argonne National Laboratory, Washington, DC, Nov. 21.

Kazanis, E., (2009), personal communication among E. Kazanis, M. Metcalf, D. Maclay, R. Bowser, and R. Baud, U.S. Department of the Interior, Minerals Management Service, Gulf of Mexico Outer Continental Shelf Region, New Orleans, LA, and C. Clark, Argonne National Laboratory, Washington, DC, Jan. 9.

Keith, K., Bauder, J. & Wheaton, J. (2003). "Frequently Asked Questions: Coal Bed Methane." Available at http://waterquality. montana.edu /docs/methane/cbmfaq.shtml#what_are_the_current_management_practic es. Accessed April 10, 2009.

Kent, M., (2008). personal communication among M. Kent, D. Asbury, and M. Abbott, Virginia Department of Mines, Minerals, and Energy, Division of Gas & Oil, Abingdon, VA, and C. Clark, Argonne National Laboratory, Washington, DC, Dec. 2.

Kentucky Geological Survey, undated, "Oil and Gas Production." Available at http://kgsmap.uky.edu/website/OGProdPlot/OGProduction. asp. Accessed May 11, 2009.

Kerr, T., (2009). personal communication between T. Kerr, Colorado Oil and Gas Conservation Commission, Denver, CO, and C. Clark, Argonne National Laboratory, Washington, DC, May 19.

Khatib, Z., & Verbeek, P. (2003). "Water to Value — Produced Water Management for Sustainable Field Development of Mature and Green Fields," *Journal of Petroleum Technology*, Jan., pp. 26–28.

Kierst, C., (2008). personal communication between C. Kierst, Utah Division of Oil, Gas, and Mining, Salt Lake City, UT, and C. Clark, Argonne National Laboratory, Washington, DC, Nov. 25.

Kierst, C., (2009). personal communication between C. Kierst, Utah Division of Oil, Gas, and Mining, Salt Lake City, UT, and C. Clark, Argonne National Laboratory, Washington, DC, April 20.

Johnston, C., Vance, G., & Ganjegunte, G. (2008). "Irrigation with Coalbed Natural Gas Co-Produced Water," *Agricultural Water Management 95*(11),1243–1252.

Lee, R., Seright, R., Hightower, M., Sattler, A., Cather, M., McPherson, B., Wrotenbery, L. Martin, D. & Whitworth, M. (2002). "Strategies for Produced Water Handling in New Mexico," presented at the 2002 Ground Water Protection Council Produced Water Conference, Colorado Springs, CO, Oct. 16–17. Available at http://www.gwpc.org/ meetings/special /PW%202002/Papers/Robert_Lee_PWC2002.pdf. Accessed April 15, 2009.

Likwartz, D., (2008), personal communication between D. Likwartz, Wyoming Oil and Gas Conservation Commission, Casper, WY, and J. Veil, Argonne National Laboratory, Washington, DC, Nov. 14.

Martin, J., (2008), personal communication among J. Martin, G. Smith, and J. Peterson, West Virginia Department of Environmental Protection, Office of Oil and Gas, Charleston, WV, and C. Clark, Argonne National Laboratory, Washington, DC, Dec. 23.

McGillivray, M., (2008), personal communication among M. McGillivray and F. Steece, South Dakota Department of Environment and Natural Resources, Division of Environmental Services, Minerals and Mining Program, Oil and Gas Section, Rapid City, SD, and C. Clark, Argonne National Laboratory, Washington, DC, Dec. 8.

McMains, S. E., (2008), personal communication among S.E. McMains, D.S. Roby, and C.P. Foerster, Alaska Oil and Gas Conservation Commission, Anchorage, AK, and C. Clark, Argonne National Laboratory, Washington, DC, Dec. 2.

NDEP (Nevada Department of Environmental Protection), (2004). *Program Description*, Bureau of Water Pollution Control, Underground Injection Control Program. Available at http://ndep.nv.gov/bwpc/docs/ nvuicprogdesc04.pdf. Accessed April 7, 2009.

Nemecek, M., (2008). personal communication between M. Nemecek, Indiana Department of Natural Resources, Division of Oil and Gas, Indianapolis, IN, and C. Clark, Argonne National Laboratory, Washington, DC, Dec. 8.

NYSDEC (New York State Department of Environmental Conservation), (2006), *New York State Oil, Gas and Mineral Resources 2006*, Division of Mineral Resources. Available at http://www.dec.ny. gov/pubs/36033.html. Accessed March 2, 2009.

NYSDEC, undated, *New York State Oil and Gas Searchable Database*, Division of Mineral Resources. Available at http://www.dec.ny.gov/cfmx/extapps/GasOil/search/production/index.cfm. Accessed March 4, 2009.

ODNR (Ohio Department of Natural Resources), (2006), *Summary of Ohio Oil and Gas Activities 2006*, Division of Mineral Resources Management. Available at http://www.ohiodnr.com/Portals/11/ publications/pdf /oilgas06.pdf. Accessed April 7, 2009.

Organek, L., (2008). personal communication between L. Organek, Michigan Department of Environmental Quality, Office of Geological Survey, Lansing, MI, and C. Clark, Argonne National Laboratory, Washington, DC, Dec. 23.

Parker, C., (2008). personal communication between C. Parker, Louisiana Office of Conservation, Oilfield Site Restoration, Baton Rouge, LA, and C. Clark, Argonne National Laboratory, Washington, DC, Dec. 15.

PDEP (Pennsylvania Department of Environmental Protection), (2008a). "*Reported Oil and Gas Production Summary for Pennsylvania*, 1985-2005," Bureau of Oil and Gas Management. Available at http://www.dcnr.state.pa.us/topogeo/oilandgas/prodreports.aspx. Accessed June 26, 2009.

PDEP, (2008b). "*Annual Production Estimations As Provided To IOGCC*," Bureau of Oil and Gas Management. Available at http://www.dep.state.pa.us/dep/deputate/minres/OILGAS/Production%20Statistics%20Estimated.htm. Accessed June 26, 2009.

Pearson, D., (2009). personal communication between D. Pearson, Arkansas Oil and Gas Commission, El Dorado, AR, and J. Veil, Argonne National Laboratory, Washington, DC, May 29.

Price, L., (2009). personal communication among L. Price, R. Land, and L. Wells, Nevada Division of Minerals, Nevada Commission on Mineral Resources, Carson City, NV, and C. Clark and J. Veil, Argonne National Laboratory, Washington, DC, Jan. 9.

Puder, M. G., & Veil, J. A. (2006), *Offsite Commercial Disposal of Oil and Gas Exploration and Production Waste: Availability, Options, and Costs*, ANL/EVS/R-06/5, prepared by the Environmental Science Division, Argonne National Laboratory, for the U.S. Department of Energy.

RRC (Railroad Commission of Texas), (2009). "Monthly Oil and Gas Production by Year, January 2003–March 2009." Available at http://www.rrc.state.tx.us/data/production/ogismcon.pdf. Accessed April 10, 2009.

RRC, undated, "Saltwater Disposal Wells Frequently Asked Questions (FAQs)."(2009). Available at http://www.rrc.state.tx.us/about/faqs/saltwaterwells.php. Accessed April 10.

Rauzi, S., (2008). personal communication between S. Rauzi, Arizona Geological Survey, Tucson, AZ, and J. Veil, Argonne National Laboratory, Washington, DC, Nov. 7.

Rauzi, S., (2009). *Annual Oil, Gas, and Helium Production in Arizona 1954-2008*, Arizona Geological Survey. Available at http://azogcc.az.gov/OG02AnnualProduction.pdf. Accessed April 6, 2009.

Shaw, H., (2009). personal communication among H. Shaw, U.S. EPA Region 10, Seattle, WA, and J. Veil, and C. Clark, Argonne National Laboratory, Washington, DC, April 29.

Sims, R., (2008). personal communication among R. Sims and R. Tarbutton, Mississippi Oil and Gas Board, Jackson, MS, and C. Clark, Argonne National Laboratory, Washington, DC, Dec. 29.

Sydow, B., (2008). personal communication among B. Sydow, Nebraska Oil and Gas Conservation Commission, Sidney, NE, and C. Clark and J. Veil, Argonne National Laboratory, Washington, DC, Dec. 5.

Taylor, D., (2008). personal communication between D. Taylor, Florida Department of Environmental Protection, Bureau of Mining & Minerals Regulation, Tallahassee, FL, and C. Clark, Argonne National Laboratory, Washington, DC, Dec. 11.

TDEC (Tennessee Department of Environment and Conservation), (2009) undated, "*Tennessee's Mineral Industry,*" *Division of Geology*. Available at http://tennessee.gov/environment/tdg/ mineralind.shtml. Accessed March 6.

U.S. EPA (U.S. Environmental Protection Agency), (2008a). "Oil and Gas Related Injection Wells (Class II)," Office of Water, Underground Injection Control Program. Available at http://www.epa.gov/ogwdw/uic/wells_class2.html. Accessed May 5, 2009.

U.S. EPA, (2008b). "*Underground Injection Control in the Southeast*," Region 4. Available at http://www.epa.gov/region4/water/uic/. Accessed April 10, 2009.

VA DMME (Virginia Department of Mines Minerals and Energy), undated, (2009) "2007 Virginia Production Summary by County." Available at http://www.dmme.virginia.gov/DGO/Production/2007 County.htm. Accessed June 12.

Veil, J., (2002). *Regulatory Issues Affecting Management of Produced Water from Coal Bed Methane Wells*, prepared for U.S. Department of Energy, Office of Fossil Energy. Available at http://www.ead.anl.gov/pub/doc/cbm-prod-water-rev902.pdf. Accessed April 10, 2009.

Veil, J., Puder, M. G., Elcock, D. & Redweik, R. J. Jr., (2004). *A White Paper Describing Produced Water from Production of Crude Oil, Natural Gas, and Coal Bed Methane.* Available at http://www.ead.anl.gov/pub/doc/ProducedWatersWP0401.pdf. Accessed May 5, 2009.

Weideman, A., (1996). "Regulation of Produced Water by the U.S. Environmental Protection Agency" in *Produced Water 2: Environmental Issues and Mitigation Technologies*, International Produced Water Symposium, M. Reed and S. Johnsen, eds., Plenum Press, New York.

WOGCC (Wyoming Oil and Gas Conservation Commission), (2009). "Download Production." Available at http://wogcc.state.wy.us/. Accessed March 9.

End Notes

[1] Argonne National Laboratory, "Produced Water Management Technology Descriptions: Fact Sheet – Discharge," Produced Water Management Information System (PWMIS), created for U.S. Department of Energy (DOE) and National Energy Technology Laboratory (NETL). Available at http://www.netl.doe.gov/technologies/pwmis/techdesc/discharge/index.html. Accessed March 30, 2009.

[2] Argonne National Laboratory, "Produced Water Management Technology Descriptions: Fact Sheet – Underground Injection for Disposal," PWMIS, created for DOE and NETL. Available at http://www.netl.doe.gov/technologies/PWMIS/techdesc/injectdisp/index.html. Accessed March 30, 2009.

[3] Argonne National Laboratory, "Produced Water Management Technology Descriptions: Fact Sheet – Underground Injection for Increasing Oil Recovery," PWMIS, created for DOE and NETL. Available at http://www.netl.doe.gov/technologies/PWMIS/techdesc/injecteor/index.html. Accessed March 30, 2009.

[4] Argonne National Laboratory, "Produced Water Management Technology Descriptions: Fact Sheet – Evaporation," PWMIS, created for DOE and NETL. Available at http://www.netl.doe.gov/technologies/PWMIS/ techdesc/evap/index.html. Accessed March 30, 2009.

[5] Argonne National Laboratory, "Produced Water Management Technology Descriptions: Fact Sheet – Offsite Commercial Disposal," PWMIS, created for DOE and NETL. Available at http://www.netl.doe.gov/technologies/PWMIS/techdesc/offsite/index.html. Accessed March 30, 2009.

[6] Argonne National Laboratory, "Produced Water Management Technology Descriptions: Fact Sheet – Agricultural Use," PWMIS, created for DOE and NETL. Available at http://www.netl.doe.gov/technologies/pwmis/techdesc/ aguse/index.html. Accessed March 31, 2009.

[7] Argonne National Laboratory, "Produced Water Management Information System," created for DOE and NETL. Available at http://www.netl.doe.gov/technologies/PWMIS/. Accessed March 31, 2009.

In: Further Explorations in the Energy-Water Nexus ISBN 978-1-62081-318-8
Editors: D. A. Hughes and B. F. Smith

Chapter 3

ENERGY-WATER NEXUS: AMOUNT OF ENERGY NEEDED TO SUPPLY, USE, AND TREAT WATER IS LOCATION-SPECIFIC AND CAN BE REDUCED BY CERTAIN TECHNOLOGIES AND APPROACHES*

United States Government Accountability Office

WHY GAO DID THIS STUDY

Providing drinking water and wastewater services are two key functions needed to support an urban lifestyle. To provide these services, energy is needed to extract, use, and treat water and wastewater. As the demand for water increases, the energy demands associated with providing water services are similarly expected to grow.

GAO was asked to describe what is known about (1) the energy needed for the urban water lifecycle and (2) technologies and approaches that could lessen the energy needed for the lifecycle and barriers that exist to their adoption. To address these issues, GAO reviewed scientific studies, government-sponsored research, and other reports and interviewed specialists

* This is an edited, reformatted and augmented version of the Highlights of GAO-11-225, a report to the Ranking Member, Committee on Science, Space, and Technology, House of Representatives, dated March 2011.

from a variety of organizations, including drinking water and wastewater utilities; federal, state, and local government offices responsible for water or energy; and relevant nonprofit groups, about the energy needed to move, use, and treat water. GAO also selected three cities—Memphis, Tennessee; San Diego, California; and Washington, D.C.—as illustrative case studies to help understand the energy demands of the lifecycle in different areas of the country.

GAO is not making any recommendations in this report. A draft was provided to the Departments of Defense, Energy (DOE), and the Interior, and the Environmental Protection Agency (EPA). DOE and EPA provided technical comments, which we incorporated as appropriate.

What GAO Found

Comprehensive data about the energy needed for each stage of the urban water lifecycle are limited. In particular, few nationwide studies have been conducted on the amount of energy used to provide drinking water and wastewater services, and these studies do not consider all stages of the lifecycle in their analysis. Specialists GAO spoke with emphasized that the energy demands of the urban water lifecycle vary by location. Considering location-specific and other key factors is necessary to assess energy needs. The specialists mentioned such factors as the topography of the area over which water is conveyed, the level and type of treatment provided, and the quality of the source water. For example, systems relying on groundwater as their source for drinking water generally use less energy than systems relying on surface water because groundwater usually contains fewer contaminants and, therefore, requires less treatment before distribution to customers.

A variety of technologies and approaches can improve the energy efficiency of drinking water and wastewater processes, but barriers exist to their adoption. Installing more efficient equipment, adopting water conservation measures, and upgrading infrastructure are among some of the approaches that can decrease energy use, according to specialists GAO spoke with and studies GAO reviewed. For example, technologies to identify potential pipeline leaks throughout water systems can reduce water loss and the energy required to pump and treat that "lost" water. However, according to specialists, adoption of technologies and approaches to improve energy efficiency may be hindered by the costs of retrofitting plants with more energy-efficient equipment and competing priorities at treatment facilities, among other barriers.

ABBREVIATIONS

DOE	U.S. Department of Energy
EPA	U.S. Environmental Protection Agency
EPRI	Electric Power Research Institute
NPDES	National Pollutant Discharge Elimination System
USGS	U.S. Geological Survey
VFD	variable frequency drive

March 23, 2011

The Honorable Eddie Bernice Johnson
Ranking Member
Committee on Science, Space, and Technology
House of Representatives

Dear Ms. Johnson:

According to the U.S. Census Bureau, in 2005, 83 percent of the U.S. population lived in metropolitan areas, up 6 percent from 2000.[1] Two key resources necessary to support an urban lifestyle are drinking water and the infrastructure necessary to treat wastewater. The average American is estimated to use about 90 gallons of water and produce 66 to 192 gallons of wastewater each day, according to the U.S. Environmental Protection Agency (EPA). As the demand for drinking water and wastewater treatment in urban areas grows, it is expected that water utilities will have to increasingly seek out alternative sources of water and treatment methods to increase the water supply, especially in areas of water scarcity where demand outpaces supply. However, treating and using these alternative sources, such as seawater, come with a cost because, in addition to other factors, they tend to be heavily energy dependent.

Providing drinking water and wastewater services to an urban environment involves extracting, moving, and treating water—referred to as the urban water lifecycle (see figure 1).[2] Energy plays a crucial role throughout this lifecycle in the following ways:

- *Drinking water processes.* Energy is needed to extract raw water from the source—such as lakes, rivers, and underground aquifers—and convey it to the drinking water treatment facility, treat the water to certain drinking water standards established under the Safe Drinking Water Act,[3] and distribute the treated drinking water to customers.
- *Customer use.* Energy is needed to circulate, pressurize, and heat water for use inside households and businesses, and for outdoor water-related uses by customers, such as watering lawns.
- *Wastewater processes.* Energy is needed to convey wastewater to treatment facilities, treat the wastewater to levels required under the Clean Water Act,[4] and discharge the treated effluent into a receiving body of water.

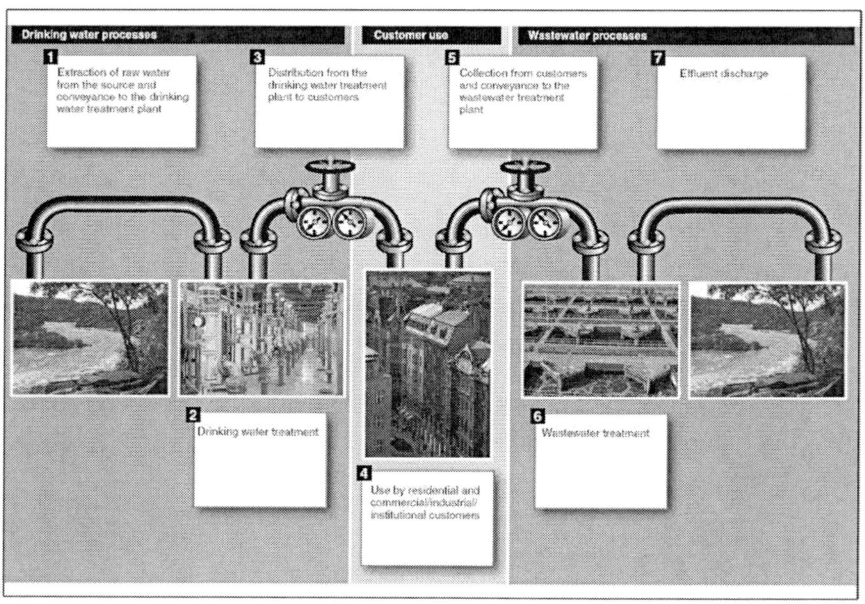

Sources: GAO analysis. Photos from left to right: GAO; US EPA Photo, Eric Vance; Art Explosion; DC Water; and GAO.

Figure 1. Key Stages of the Urban Water Lifecycle.

As urban populations increase and the demand for water grows, the energy needed for the urban water lifecycle is also expected to grow. In this context, you asked us to review the energy needs of providing drinking water and wastewater treatment services to urban users. Specifically, the objectives of this review were to describe what is known about (1) the energy needed for each stage of the urban water lifecycle and (2) technologies and approaches that could lessen the energy needs of the urban water lifecycle, as well as any identified barriers that exist to their adoption.

To address both of these objectives, we conducted a systematic review of studies and other documents that examine the energy required to extract, move, use, and treat water, including peer-reviewed scientific and industry periodicals, government-sponsored research, and reports from nongovernmental research organizations. We also selected a nonprobability sample of three cities to examine in greater depth and better understand regional and local differences related to urban water lifecycles: Memphis, Tennessee; San Diego, California; and Washington, D.C. We chose these cities as illustrative case studies based on criteria such as type of water source; water availability; type of wastewater system; unique characteristics, such as potential to treat seawater to help meet drinking water demands; and economic factors, such as energy costs. While the information derived from our analysis of these cities cannot be generalized to all U.S. cities, these examples provide valuable insights regarding the complexities of assessing the energy needs for the urban water lifecycle. For each of these case studies, we analyzed documentation from, and conducted interviews with, a wide and diverse range of specialists from organizations involved in all stages of the urban water lifecycle. These organizations included drinking water and wastewater treatment facilities, and state and local agencies responsible for water or energy.[5]

In addition to specialists associated with the illustrative case studies, we interviewed a range of other knowledgeable individuals whom we identified as having expertise related to the energy needs of all stages of the urban water lifecycle throughout the United States. We selected these specialists using an iterative process, soliciting additional names from each person we interviewed. From among those identified, we interviewed specialists who could provide us with a broad range of perspectives on the energy needs of the urban water lifecycle. We also interviewed specialists whom we identified during our systematic review of studies who have analyzed (1) the energy needed in one or more stages of the water lifecycle at the national or local level or (2) techniques available to reduce the energy demands for water. These specialists represented a variety of organizations, including drinking water and wastewater treatment facilities; state and local government offices responsible for water or energy; officials from EPA and researchers from some of the U.S. Department of Energy's (DOE) national laboratories, such as Sandia National Laboratory; university researchers; water and energy industry representatives from groups such as the American Water Works Association and the Water Research Foundation; and relevant nongovernmental organizations, such as the Pacific Institute, a nonpartisan research institute that works to advance environmental protection, economic development, and social equity. The specialists also included individuals with knowledge of the energy demands for water in other states, including Arizona, Colorado, Florida, New York, and Wisconsin, to provide a better understanding of water and energy issues in other regions around the United States.

We also interviewed other federal agency officials, scientists, and researchers and analyzed data and information from federal agencies that have responsibilities related to the energy needs of the urban water lifecycle—the Department of Defense's U.S. Army Corps of Engineers, DOE, the Department of the Interior's U.S. Geological Survey (USGS) and Bureau of Reclamation, EPA, and the National Science Foundation. We performed our work from January 2010 to January 2011 in accordance with all sections of GAO's Quality Assurance Framework that are relevant to our objectives. The framework requires that we plan and perform the engagement to obtain sufficient and appropriate evidence to meet our stated objectives and to discuss any limitations in our work. We believe that the information and data obtained, and the analysis conducted, provide a reasonable basis for any findings and conclusions in this product.

BACKGROUND

According to EPA, about 52,000 community water systems use energy to treat and deliver drinking water to over 290 million Americans.[6] In a typical drinking water treatment plant, large debris and contaminants are physically removed from the raw water using screens (see figure 2). Next, dirt and other particles suspended in the water are removed through the addition of alum and other chemicals during the processes of coagulation and sedimentation. After these particles have separated out, the water passes through filters made of layers of materials such as sand, gravel, and charcoal to remove even smaller particles. At this point, the water is stored in a closed tank or reservoir, allowing time for disinfection which kills many disease-carrying organisms. The distribution infrastructure consists of pumps, pipes, tanks, valves, hydrants, and meters that support delivery of water to the customer and control flow and water pressure.

Once water is delivered, residential consumers use it for a variety of purposes, including for drinking; bathing; preparing food; washing clothes and dishes; and flushing toilets, which can represent the single largest use of water inside the home. Energy is needed to accomplish many of these activities. For example, energy is used in homes to filter and soften water and to heat it for use in certain appliances, which accounts for 12.5 percent of a typical household's energy use, according to DOE. In addition to residential water users, commercial, industrial, and institutional customers use energy for water-related purposes. For example, energy is used to produce hot water and steam for heating buildings, to cool water for air conditioning buildings, and to generate hot water needed to manufacture or process materials, such as food and paper.

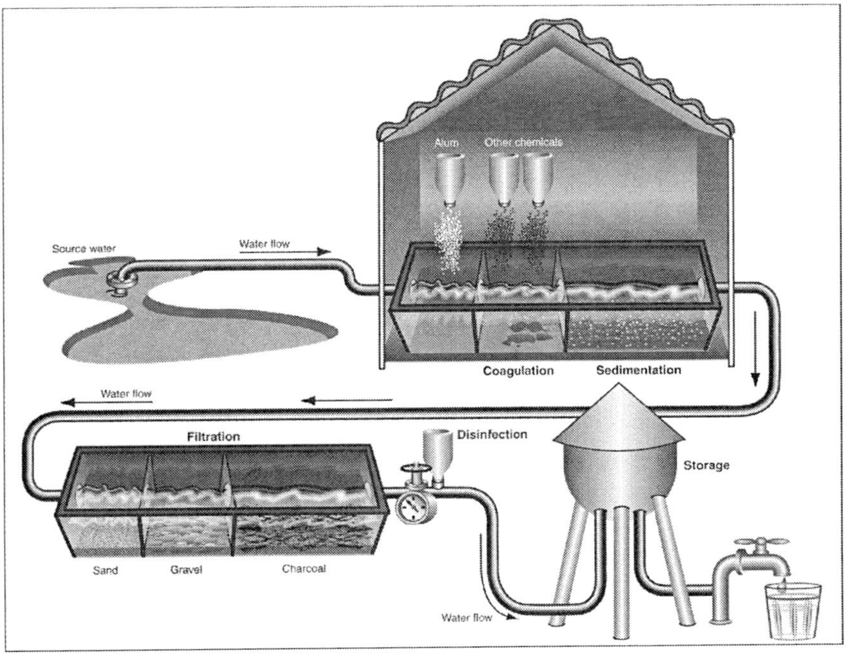

Source: GAO analysis.

Figure 2. Typical Drinking Water Treatment process.

After water is used by customers, energy is needed to collect and treat wastewater, and to discharge effluent into a water body. Wastewater service is provided to more than 220 million Americans by about 15,000 municipal wastewater treatment facilities.[7] During a typical wastewater treatment process, solid materials, such as sand and grit, organic matter from sewage, and other pollutants, are removed before the treated effluent is discharged to surface waters. Systems for collecting, treating, and disposing of municipal wastewater vary widely in terms of the equipment and processes used, and wastewater may go through as many as three treatment stages—primary, secondary, and advanced treatment—before water is discharged (see figure 3).

- *Preliminary and primary treatment.* As wastewater enters a treatment facility, it is screened to remove large debris and then passes through a grit removal system to separate out smaller particulate matter. After preliminary screening and settling, primary treatment removes solids from the wastewater through sedimentation. Solids removed during the treatment process may be further treated and used for other applications, such as fertilizer; incinerated; or disposed of in landfills.
- *Secondary treatment.* After primary treatment, the wastewater undergoes secondary treatment to remove organic matter and suspended solids through physical and biological treatment processes. Activated sludge is the most commonly used biological treatment process in secondary treatment of wastewater. This process relies on micro-organisms to break down organic matter in the wastewater. More specifically, aeration—whereby blowers or diffusers inject oxygen into the wastewater—enables the micro-organisms to digest the organic matter. After being pumped into an aeration tank to allow time for digestion, the wastewater is next pumped to a secondary settling tank for removal of digested material. After secondary settling, the effluent either is disinfected and discharged into a water body, or it undergoes advanced treatment.
- *Advanced treatment.* Most wastewater goes through at least secondary treatment. However, before treated wastewater can be released in some receiving waters, it may need to be further treated to reduce its effect on water quality and aquatic life after discharge. Over 30 percent of wastewater treatment facilities provide this kind of advanced treatment, which can remove additional contaminants.

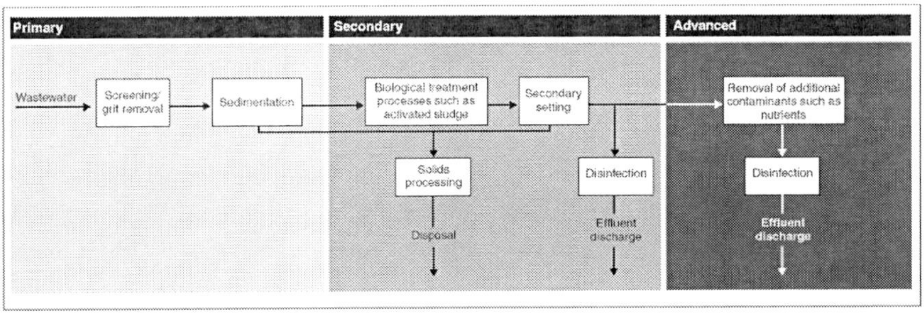

Source: GAO analysis.

Figure 3. Typical Wastewater Treatment Process.

Two key pieces of federal legislation—the Safe Drinking Water Act and the Clean Water Act—govern the treatment of drinking water and wastewater. Each municipality or water utility generally may choose amongst technologies for achieving a given standard. Under the Safe Drinking Water Act, EPA has established National Primary Drinking Water Standards for specified contaminants and has the authority to regulate additional contaminants that the agency determines may have adverse health effects, are likely to be present in public water supplies, and for which regulation presents a meaningful opportunity for health risk reduction. EPA's regulations establish a limit, or maximum contaminant level, for specific contaminants and require water systems to test the water periodically to determine if the quality is acceptable.[8] EPA has regulations in place for 89 contaminants, including disinfectants, byproducts of disinfectants, and microbial contaminants, but has not issued a regulation under the Safe Drinking Water Act for a new contaminant since 2000.

The Clean Water Act governs the discharge of pollutants into the waters of the United States, including the treatment of wastewater discharged from publicly owned treatment facilities. Specifically, industrial and municipal wastewater treatment facilities must comply with the National Pollutant Discharge Elimination System (NPDES) permits that control pollutants that facilities may discharge into the nation's surface waters. The act requires that municipal wastewater treatment plants provide a minimum of secondary treatment prior to discharge. In some cases, modification of secondary treatment requirements may occur, however, for discharges into marine waters under certain conditions. For example, the discharge may not interfere with that water quality which assures protection of public water supplies and the protection and propagation of a balanced, indigenous population of shellfish, fish, and wildlife and allows recreational activities on the water. In 2000, Congress amended the Clean Water Act to require permits for discharges from combined sewers—sewers that transport both wastewater and stormwater to the municipal wastewater treatment plant—to conform with EPA's Combined Sewer Overflow Control Policy, which requires systems to demonstrate implementation of certain minimum pollution control practices.[9] Combined sewers may overflow when there is heavy precipitation or snowmelt, resulting in the discharge of raw sewage and other pollutants into receiving water bodies.

Comprehensive Data about the Energy Needed for the Urban Water Lifecycle are Limited, However Energy Needs are Influenced by Location-Specific Factors

Comprehensive data about the energy needed for each stage of the urban water lifecycle are limited, and few nationwide studies have been conducted on the amount of energy used to provide drinking water and wastewater treatment services to urban users. However, specialists with whom we spoke emphasized that the energy demands of the urban water lifecycle vary by location; therefore, consideration of location-specific and other factors is key to assessing the energy needs of the urban water lifecycle. These factors include the source and quality of the water, the topography of the area over which water is conveyed and the distance of conveyance, and the level and type of treatment required.

Comprehensive Data on the Energy Needed to Support the Urban Water Lifecycle are Limited

Providing a reliable and comprehensive estimate of the total energy requirements for moving, treating, and using water in urban areas is difficult, in part, because comprehensive data on the energy demands of the urban water lifecycle are limited and few nationwide studies have been conducted to quantify the amount of energy used throughout the lifecycle. Two studies most often cited by the specialists we spoke with were conducted by the Electric Power Research Institute (EPRI) on the energy needs of the urban water lifecycle. These studies concluded that 3 to 4 percent of the nation's electricity is used to move and treat drinking water and wastewater. While some specialists noted that these studies provide reasonable estimates of the energy demands of the urban water lifecycle, other specialists raised a number of concerns with the studies. In particular, according to several specialists, the EPRI studies are outdated. The first study dates back to 1996, and the more recent study was conducted in 2002 but relied on projections of future water use based on statistics compiled in 2000. Some specialists also told us these studies do not reflect the treatment processes that have been implemented over the last decade, which have increased the amount of energy needed to treat water. In addition, the studies do not include all stages of the urban water lifecycle—specifically, they omit energy used by customers. Because they exclude end use, the EPRI studies underestimate the energy demands of the entire lifecycle because customer end use, including use by residential customers, can be the most energy-intensive stage of the entire lifecycle, according to some specialists we spoke with and studies that we reviewed. Some specialists also added that the studies underestimate total energy demands because they include only electricity, excluding other fuel types that can be used throughout the lifecycle. For example, the studies do not assess the use of natural gas, which can be a primary energy source at wastewater treatment plants for certain processes. Furthermore, some specialists explained that the studies do not use actual measured data, relying instead on previously published estimates of energy used for portions of the water lifecycle.

Furthermore, some specialists noted that there are limited data on the amount of energy associated with customer water use. Federal agencies like DOE's Energy Information Administration collect some data on energy used to heat water in residences and in the commercial sector, but these data are reported on a national level and do not allow for analysis at the local level. In addition, data needed to get a full picture of the energy needs for water in an urban setting may not be readily available at the local level. Specifically, water utilities may not have detailed data on their facilities' energy use, may not have conducted audits to understand how their facilities use energy, or may be reluctant to share data, according to specialists we spoke with.

Consideration of Location-Specific and Other Factors Is Key to Assessing the Energy Needs of the Urban Water Lifecycle

Many of the specialists told us that efforts to assess the energy needs of the urban water lifecycle on a national scale can be difficult, and the majority of the specialists we spoke with emphasized that to obtain a more accurate picture, one needs to consider location-specific and other factors that influence energy use. The specialists identified the following as key factors that must be considered for such an assessment.

Type of water source. Drinking water systems that rely on surface water are often designed to take advantage of gravity and use little to no energy to extract water from the source and convey it to the treatment facility. In contrast, systems that rely on groundwater require more energy for extraction because water must be pumped to the surface from underground aquifers, especially if they rely on deep underground aquifers. For example, Washington, D.C., which relies on surface water, withdraws its water from two locations—Great Falls Dam and Little Falls Dam—on the Potomac River. Most of the water is withdrawn at Great Falls Dam and conveyed via gravity to the treatment plant, using little energy during the extraction and conveyance process. In contrast, extraction of water is an energy-intensive process for Memphis, which relies on groundwater that is extracted from over 160 wells that draw water from aquifers including the Memphis Sand Aquifer, located 500 to 600 feet below ground.

Quality of water to be treated. The quality of water also impacts the amount of energy needed for treatment, with higher-quality water containing fewer contaminants and, therefore, requiring less treatment than lower quality water. For example, treating groundwater generally uses less energy than treating surface water because groundwater is typically of higher quality than surface water. As a result, cities that rely on groundwater as the source for their drinking water, such as Memphis, generally use less energy for treatment than cities that rely on surface water, such as Washington, D.C. However, the type of contaminants in water can also affect the energy required for treatment. For example, as one specialist noted, if groundwater contains arsenic, treating this type of contamination can require the use of more energy-intensive treatment technologies than treating surface water that is extracted from a protected watershed or clean snowmelt.

Topography and distance. Pumping water is one of the most energy-intensive aspects of the urban water lifecycle, accounting for 80 to 90 percent of the energy used to supply drinking water in some systems, and most of this energy is used to distribute water to customers. The energy demand of pumping is affected by the topography over which the water must be moved and the distance the water must travel to treatment plants after extraction and to customers after treatment.[10] For example, San Diego gets a large amount of its water from northern California. Transporting this imported water to southern California is energy intensive because the water must be conveyed hundreds of miles and lifted 2,000 feet over the Tehachapi Mountains. Furthermore, because of the hilly terrain in some parts of the city and the great expanse over which the customers are distributed, additional energy is needed to pump water from the treatment plants to consumers.

Condition of water system. The age of a system and the condition of its pipes and equipment can also impact the energy demands of providing drinking water and wastewater treatment services. Specifically, older systems can be less energy efficient if the equipment and infrastructure have not been properly maintained. The American Society of Civil Engineers recently evaluated America's drinking water and wastewater infrastructure and assigned both systems a grade of a D-. The assessment noted that these systems contain facilities that are nearing the end of their useful lives and need upgrades to meet future regulatory requirements. In addition, the condition of pipelines also has energy implications. According to some specialists we spoke with, up to 50 percent of water is lost through leaking pipes, which results in a loss of the energy that was used to extract, convey, and treat the water. Furthermore, if pipelines are not routinely cleaned, blockages can lead to friction in the pipes, requiring additional energy to push water through these pipes.

Required treatment level. Energy needed for drinking water and wastewater treatment is affected by the treatment levels required to meet existing water quality standards, with each additional treatment level increasing energy demands. In the case of wastewater treatment, characteristics of the water body into which treated effluent is discharged can impact the required level of treatment. For example, San Diego officials told us the city's wastewater treatment facility has been granted a modified permit by EPA. According to these officials, this permit allows San Diego to treat its wastewater only to an advanced primary level in part because years of ocean monitoring have shown that the plant discharges have no negative impact to the Pacific Ocean.[11] If the city had to treat its wastewater to secondary treatment levels, city officials estimate that its energy usage would increase six to nine times as a result of having to use more energy-intensive technologies to meet these higher standards.

Type of treatment process. The type of treatment process used at drinking water and wastewater facilities also influences the energy demands of providing drinking water and wastewater services to urban users. For example, treatment plants that use the activated sludge process for secondary treatment use more energy than plants that use other processes, such as trickling filters or lagoon systems.[12] The activated sludge process can account for 70 percent of a wastewater treatment plant's energy consumption because of the energy needed to power the blowers that pump oxygen into the wastewater to sustain the microorganisms. Furthermore, according to many of the specialists we spoke with, a number of the new technologies used in drinking water treatment plants are more energy intensive than traditional treatment technologies. For example, some treatment plants are installing ultraviolet light disinfection processes that are more energy intensive, accounting for 10 to15 percent of a plant's total energy use, than traditional disinfection with chlorine.[13] Other energy-intensive technologies that are increasing energy demands for water treatment include filtration using membranes and ozonation, a process that destroys bacteria and other micro-organisms through an infusion of ozone.

Water use and type of customer. Characteristics related to customer water use, such as how and where water is consumed, can also influence the amount of energy needed to provide water and wastewater services to urban users, according to specialists we spoke with. Large amounts of household energy are consumed by heating water for showering, dishwashing, and other uses. These uses would require more energy than other household uses, such as flushing toilets. In addition, some specialists told us that where the water is used influences the amount of energy consumed. For example, water used in tall apartment buildings or skyscrapers requires energy-intensive pumps to move the water to the top floors. Furthermore, according to some specialists we spoke with, the type of customer, such as whether the customer is residential or industrial, affects the energy demands of providing water and wastewater services. For example, Memphis has two wastewater treatment plants, one of which is located in an industrial section of the city and receives a higher percentage of its wastewater from industrial sources than the other facility, which receives a higher percentage of its wastewater from residential sources. Because the industrial wastewater contains increased levels of organic contaminants and thus requires more energy for treatment, the two facilities consume different amounts of energy on a per-gallon basis.

Water availability. As current water supplies diminish, some cities, especially those in areas that are already water stressed, are moving toward alternative water supply sources that will require more energy for treatment than processes used for surface water and groundwater. For example, to help meet future demands for water and reduce dependence on imported water supplies in San Diego, the region is pursuing energy-intensive seawater desalination, which can be 5 to 10 times more energy intensive than conventional processes to treat surface water and groundwater. Other areas, such as Tucson, Arizona, that do not have ready access to seawater are pursuing desalination of brackish groundwater—water that is less saline than seawater but that contains higher saline levels than found in freshwater. Although treating brackish water is less energy intensive than seawater desalination, it still can use two to three times more energy than conventional water treatment processes for freshwater supplies. Furthermore, San Diego is studying the viability of treating a portion of its reclaimed water—wastewater effluent that is treated to an advanced level and suitable for nonpotable water applications such as irrigation—for potable water use. To implement such a system, San Diego would need to add energy-intensive advanced treatment processes to its current wastewater treatment system. However, because this additional energy use would offset the energy demands for imported water, city officials told us the project is expected to result in a net reduction in San Diego's energy profile. Using reclaimed water can also increase energy demands for pumping, depending on the design of the existing wastewater system. That is, many wastewater collection systems were designed with treatment plants located in low elevation areas to take advantage of gravity in conveying the wastewater to the plant. However, if wastewater is recycled, energy could be needed to pump this water against the flow of gravity into the distribution system, but such increases may actually be less energy intensive than reliance on imported water.

Future regulatory changes. To address growing concerns about emerging contaminants and nutrients in the nation's water bodies, according to many specialists, additional or more stringent regulatory standards could increase the energy demands of treatment processes in the future. Specifically, any more stringent standards that are promulgated would most likely require additional levels of treatment, and energy-intensive technologies, such as ozonation and membrane filtration, may be necessary to meet such new standards. More stringent regulations in the future could also increase energy demands even for facilities that have already implemented such technologies. For example, according to officials of the Washington, D.C., wastewater treatment plant, while the facility already must meet the nation's most stringent permit requirements and uses advanced treatment processes, stricter standards are expected to increase the plant's energy demands, in part, because new energy-intensive technologies may need to be added to the plant's treatment process. Regulatory changes could also increase energy demands at other stages of the urban water lifecycle. For example, higher standards for effluent discharge from wastewater treatment plants could increase the energy required for treatment. Furthermore, stricter water quality standards for receiving waters could necessitate more plants to employ advanced treatment standards, resulting in increased energy use for the additional treatment or to pump effluent farther away to other waters.

Complexity of water systems. In addition to location-specific factors, the complexity of some urban water systems can make assessing the energy demands of the urban water lifecycle challenging. For example, some urban water systems like San Diego's are highly complex, involving a number of different entities that have responsibility for different parts of the system. Specifically, the City of San Diego currently imports 85 to 90 percent of its water from the Colorado River and northern California. In addition, the city's regional drinking water, wastewater, and recycled water systems are managed by a number of different organizations responsible for conveying drinking water, wastewater, and recycled water to multiple treatment facilities with over 160 pumping stations spread over 400 square miles within the City of San Diego's service territory alone. As a result, collecting consistent data on energy use from each of these organizations is challenging, according to San Diego water officials we spoke with.

CERTAIN TECHNOLOGIES AND APPROACHES CAN REDUCE ENERGY USE, BUT BARRIERS COULD IMPEDE THEIR ADOPTION

Specialists we spoke with and studies we reviewed identified a variety of technologies and approaches that can improve the energy efficiency of drinking water and wastewater processes associated with the urban water lifecycle, and determining the appropriate solution depends on the circumstances of a particular system. However, adoption of these technologies and approaches may be hindered by costs; inaccurate water pricing; barriers associated with operational factors, such as limited staffing levels at water utilities; competing priorities at drinking water and wastewater facilities; and lack of public awareness about the energy demands of the urban water lifecycle.

Certain Technologies and Approaches Can Reduce the Energy Use of the Urban Water Lifecycle

Several key technologies and approaches are currently available that can improve the energy efficiency of drinking water and wastewater processes, but determining the most appropriate solution depends on the circumstances of a particular system and requires an understanding of the system's current energy use. Many studies that we reviewed and specialists we spoke with identified process optimization, equipment and infrastructure upgrades, water conservation, and improved energy management as approaches that can help reduce the energy demands for water. In addition, the increased use of renewable energy could offset the energy purchased by water utilities from energy providers.

Process Optimization

According to some studies we reviewed, energy consumption by water and wastewater utilities can comprise 30 to 50 percent or more of a municipality's energy bill. Optimizing drinking water and wastewater system processes, including energy-intensive operations like pumping and aeration, was identified in many studies that we reviewed as an approach to reducing the energy demands of the urban water lifecycle. Implementing monitoring and control systems and modifying pumping and aeration operations are some ways to reduce energy use through process optimization.

- *Implementing monitoring and control systems.* Monitoring and control systems, also known as supervisory control and data acquisition systems, can be used to optimize drinking water and wastewater operations. Such systems provide a central location for monitoring and controlling energy-consuming devices and equipment, which provides plant operators with the ability to schedule operations or automatically start and stop devices and equipment to manage energy consumption more effectively and improve overall operations.
- *Modifying pumping operations.* A variety of modifications could increase the efficiency of pumping systems. For example, operating constant speed pumps as near as possible to their most efficient speed, using higher efficiency pumps as opposed to lower efficiency pumps, and operating multiple smaller pumps rather than a few large pumps to better match pumping needs can help maximize pumping efficiency. In addition, using devices to monitor and control pump speeds—known as variable frequency drives (VFD)—may allow facility operators to accommodate variations in water flows by running pumps at lower speeds and drawing less energy when water flows are low. Potential energy savings from the use of VFDs can range from 5 to 50 percent or more, according to studies we reviewed. However, these studies and some specialists we spoke with also noted that VFDs are not necessarily well suited for all applications—such as when flow is relatively constant—and that potential benefits of VFDs should be evaluated based on system characteristics, such as pump size and variability of flow.
- *Modifying aeration operations.* According to many studies we reviewed and specialists we spoke with, aeration in wastewater treatment consumes a significant amount of energy, and systems can be reconfigured and better controlled to improve energy efficiency. Specifically, blowers and mechanical aerators are typically powered

by a large motor, and installing variable controls on blowers to enable operators to better match aeration with oxygen requirements can reduce energy demands. Likewise several studies noted that dissolved oxygen control systems can be used to match oxygen supply with demand by monitoring the concentration of dissolved oxygen in the wastewater and adjusting the blower system or mechanical aerator speed accordingly. In addition, probes can be installed to monitor dissolved oxygen levels within the wastewater and signal operators when the system may need adjustment.

Equipment and Infrastructure Improvements

According to many studies and specialists we spoke with, installing more efficient equipment—motors, pumps, blowers, and diffusers—for energy-intensive processes such as aeration and pumping can reduce energy use. In addition, ensuring the proper sizing and maintenance of equipment and infrastructure can improve energy efficiency.

- *Upgrading equipment.* Replacing less efficient equipment with more energy-efficient equipment can reduce energy use. For example, installing more efficient motors could reduce energy use by 5 to 30 percent, according to studies we reviewed. In addition, blower and diffuser technologies, including high-speed "turbo" blowers and fine or ultra-fine bubble diffusers, could decrease the energy demands of aeration. High-speed turbo blowers use less energy than other blower types, although, because these blowers are a new technology and relatively few are in use, efficiency claims are not yet well documented, according to a 2010 EPA report.[14] Energy-saving estimates for fine bubble diffusers, which have higher oxygen transfer efficiencies than coarse bubble diffusers, range from 9 to 50 percent or more, but some specialists and studies expressed concerns about maintenance requirements as well as the durability of this technology.
- *Right-sizing equipment.* Many wastewater treatment systems were designed to handle greater capacity in the future because of anticipated population growth. However, this growth has not always occurred and, as a result, existing equipment may be oversized and consume more energy than is needed to treat current flows, according to some specialists we spoke with. Proper sizing and selection of

pumping and aeration equipment to more closely match system needs can help maximize efficiency. For example, in Washington, D.C., the operators of the wastewater treatment plant replaced a 75-horsepower motor with a 10-horsepower motor in one facility to better meet actual energy demands.

- *Improving maintenance and leak detection technology*. Periodic inspections to assess pump performance and the need for replacement or maintenance of electrical systems and motors can increase the energy efficiency of the overall system, according to studies we reviewed. In addition, leak detection technologies can identify leaks throughout water systems, thereby reducing water loss and the related energy required to pump and treat that "lost" water. For example, acoustic leak detection systems use sensors to monitor for sounds that may indicate potential leaks and relay the data back to a central control room, which helps water utility staff identify actual leaks and schedule maintenance accordingly. The San Diego County Water Authority, which provides water to San Diego and other areas in southern California, has fiber optic lines in place to monitor its pipeline 24 hours a day to detect evidence of leaks.

Water Conservation and Efficiency

Many studies we reviewed and specialists we spoke with also identified water conservation as an approach to reducing the energy needed for the urban water lifecycle. Several studies noted that decreased customer water use could directly translate into energy savings. Furthermore, water conservation also reduces the amount of energy used to convey, treat, and distribute drinking water to the customers. Studies we reviewed and specialists we spoke with identified a variety of tools that utilities can use to promote water conservation, including enhanced metering, increased water prices, public education, and incentives to install water-efficient appliances. For example, San Diego is implementing advanced metering tools to better manage its system and to provide real-time information to customers regarding their water use in order to help them make choices that conserve water. In addition, EPA has developed water efficiency and performance criteria for several product and program categories through WaterSense, a federal water efficiency program.[15]

Energy Management

While many technologies and approaches have been identified to reduce the energy demands for water, determining the most appropriate solution depends on the circumstances of a particular system—including the type of facilities and treatment processes in place—and requires an understanding of current energy use. Several studies we reviewed identified improved energy management, including conducting energy audits of treatment facilities or systems, as a necessary first step to reducing energy demands. Specifically, specialists told us that by providing utility managers with information about their facilities' energy use, energy audits can help managers identify opportunities to change plant operations in ways that will save energy. For example, the energy supplier for one wastewater treatment plant in Memphis conducted an energy audit of the blower system, which used about 75 percent of the plant's total energy. As a result of this audit, operators changed their practices to run blowers at the lowest levels possible while still ensuring they continued to meet the effluent discharge standards required by the plant's permits.

Similarly, in 2000, San Diego established an in-house energy management program, which includes an audit team that looks for technologies and approaches to lessen the energy demands of the city's drinking water and wastewater systems. The team studies the efficiency of existing equipment and treatment processes and considers upgrading or replacing equipment with available energy-efficient technologies. For example, the energy audit team identified over a dozen energy conservation measures that could be applied to reduce energy consumption at two of the city's sewer pump stations, including installing timers to turn off lighting and upgrading, resizing, and replacing motors and blowers.

In addition, EPA's Energy Star program provides energy management tools and strategies to support the successful implementation of energy management programs. Officials told us that EPA also works with municipal drinking water and wastewater utilities to provide information on potential energy efficiency opportunities. EPA's online benchmarking tool, known as the Portfolio Manager, offers wastewater treatment plant managers the opportunity to compare the energy use of their plants with that of other plants using the EPA energy performance rating system. EPA has also published a variety of educational materials for drinking water and wastewater utilities to help identify, implement, measure, and improve energy efficiency and renewable energy opportunities.

Other Approaches

Specialists we spoke with and studies we reviewed identified two additional approaches for reducing the energy required to treat and distribute water: improving advanced treatment technologies and redesigning a city or region's water system.

- *Improving advanced treatment technologies.* According to EPA officials, and as previously noted by specialists, improving energy-intensive advanced treatment technologies—such as ultraviolet disinfection, ozone, and membrane technologies—is important because plants are increasingly using them. For example, the use of membrane materials that require less pressure to push water through to remove contaminants could decrease the energy demands of that technology. In addition, some specialists we spoke with told us that newer technologies are being developed, such as forward osmosis, that may offer alternative treatment approaches that are more efficient than the technologies currently used for desalination. Several specialists told us the federal government should conduct additional research to understand and improve the energy efficiency of water supply, treatment, and water use—for example, by conducting more research on energy-efficient desalination technologies.
- *Redesigning water systems.* Some specialists noted that redesigning water systems in ways that better integrate drinking water, wastewater, and stormwater management could improve the energy efficiency of water systems overall. Decentralizing treatment systems, implementing approaches to better manage stormwater, reusing wastewater, and using less energy-intensive processes for biological treatment can help reduce energy needed for providing drinking water and wastewater services. For example, current water systems primarily rely on a few plants with large capacities to treat drinking water and wastewater. Some specialists told us that systems could be redesigned to incorporate more treatment plants with smaller capacities and to locate these plants closer to the point of water use by

customers, thereby reducing some of the energy required for pumping to the treatment site. In addition, some specialists identified improvements in stormwater management through strategies such as low-impact development—which involve land use planning and design to better manage stormwater—as a way to reduce the energy required for treatment. For example, by decreasing stormwater infiltration into some wastewater systems through low-impact development activities such as the capture and use of rainwater, flows into treatment plants would also be reduced, thereby decreasing the energy needed for treatment. In addition, reusing wastewater for purposes that may not require potable water, such as industrial processes or landscaping, may reduce overall energy use by decreasing energy used currently to pump, treat, and distribute potable water to these customers, according to some studies we reviewed. However, the potential for energy savings from reuse depends on the energy intensity of a given system's water supply as well as the level of treatment needed for potential uses. Furthermore, some studies we reviewed and specialists we spoke with noted that relying more on biological treatment processes that do not require aeration, such as using lagoons or trickling filters, may be an option to reduce energy demands. However, these approaches may be limited by available space in urban areas and therefore may not be applicable everywhere.

Renewable Energy

Many studies we reviewed and specialists we spoke with stated that drinking water and wastewater utilities could adopt renewable energy projects to reduce energy purchased from energy providers. Renewable energy projects may include solar, wind, and hydroelectric power as well as the recovery and use of biogas from wastewater treatment processes.[16] In addition, some studies we reviewed and specialists we spoke with identified hydro turbines as an option for recovering energy in the distribution system. For example, water systems with changes in topography that have pressure-reducing valves in place can install turbines that generate electricity as water flows past. This energy could then be recovered for use in powering equipment.

The city of San Diego has adopted a variety of renewable energy projects to power its drinking water and wastewater treatment operations. For example, the city installed a 945-kilowatt solar power system at the Otay Water Treatment Plant that produces enough electricity to meet the power needs of the plant's pumping operation (see figure 4). In addition, at the city's Point

Loma Wastewater Treatment Plant, both methane and hydroelectric power are recovered from wastewater processes. The plant uses digestion processes to treat organic solids resulting from its wastewater treatment processes. Methane, a by-product of the digestion process, is removed from the digesters and used to power two engines that supply all of the plant's energy needs, making it energy self-sufficient. In addition, the plant recovers hydroelectric power from the treated effluent that it discharges into the ocean. The effluent drops 90 feet from the wastewater treatment plant to the ocean, powering a 1,350 kilowatt hydroelectric plant. The city can sell any excess energy produced by the plant back to the electric utility.

While renewable energy projects have the primary benefit of reducing the energy needed by water treatment facilities from outside providers, such projects could also reduce overall energy use. For example, solar power systems co-located at treatment facilities in San Diego may result in the offset of slightly more electricity than they produce, since electricity generated by the energy provider off-site and transferred over a greater distance results in some loss of energy during transmission.

Source: San Diego Public Utilities Department.

Figure 4. Solar Panels at San Diego's Otay Water Treatment Plant.

Key Barriers Could Impede Adoption of Technologies and Approaches

Specialists we spoke with identified a number of key barriers to adopting the available technologies and approaches that could reduce the energy demands of the urban water lifecycle. These barriers fall into five categories: (1) costs associated with these technologies, (2) inaccurate water pricing, (3) barriers associated with how water utilities operate, (4) competing priorities at drinking water and wastewater facilities, and (5) the lack of public awareness about the energy demands of the urban water lifecycle.

Costs Associated with Energy-Saving Improvements

Energy-saving technologies may lessen the energy demands of the urban water lifecycle, but such improvements are often expensive to adopt. Many specialists told us that, as a result, utilities may not be able to justify the costs necessary to install energy-efficient equipment. For example, some specialists told us that upgrading to VFDs, higher-efficiency pumps, and ultra-fine bubble diffusers may lessen a water facility's energy demands, but the costs of installing these technologies can be prohibitive for some systems, and it can take years to realize the full energy-saving benefits. As a result, some utility operators may choose to wait until there is an immediate need to upgrade equipment because the costs can be justified more easily at that point. Similarly, some specialists told us that the cost of installing renewable energy projects, such as solar panels, can be a barrier to adoption for some treatment facilities. According to an energy specialist we spoke with, it may take over 30 years to fully realize the cost savings from such projects. However, a DOE official noted that while expensive in the past, the cost of solar panels has been decreasing in recent years. Furthermore, installing energy-efficient equipment and infrastructure upgrades, such as replacing leaking pipelines, can be particularly challenging for smaller water utilities because they often compete for limited funds against other municipal services, such as fire and police protection. In addition, in areas where energy costs are low, there may be little incentive for water utility operators to implement capital-intensive practices to save energy. To help overcome the barriers associated with the costs of upgrading facilities, some specialists told us that utilities should conduct cost analyses to account for the total savings incurred over the life of the energy-saving projects, not just focus on the short-term returns on investment. Some specialists we spoke with also suggested that utilities should take advantage of the federal funding available through the Drinking Water and Clean Water

State Revolving Fund programs, which can be used to fund a variety of projects that improve water and energy efficiency. These programs provide financial assistance for drinking water and wastewater infrastructure projects, respectively, and for certain other purposes, such as installing water meters, installing or retrofitting water-efficient devices, and promoting water conservation. In addition, the American Reinvestment and Recovery Act of 2009 and EPA's fiscal year 2010 appropriation encourage states to use a portion of those funds for such energy and water efficiency projects.[17]

Inaccurate Water Pricing

According to many specialists with whom we spoke, the true cost of water is often not reflected in rates customers are charged. Specifically, specialists told us that water subsidies have kept water rates artificially low and do not reflect the actual cost, including energy costs, of pumping, treating, and moving drinking water and wastewater. The effect of this situation is two-fold. First, there may be little incentive for customers to use water more efficiently if they are not paying the true cost of it. Second, because these reduced water rates generally do not cover the actual costs incurred by drinking water and wastewater facilities, some utilities do not generate enough revenue to implement upgrades that could lessen their facilities' energy demands. Some specialists noted, however, that rate increases are not a politically popular approach and may be met with public and political resistance.

Operational Challenges

Other barriers to adopting energy-reducing technologies and approaches are operational in nature. Specifically, specialists we spoke with noted a number of such challenges, including utilities not having staff with adequate knowledge about technologies and access to energy-use data, reluctance to change, and lack of coordination between water and energy utilities. For example, several specialists told us that smaller utilities lack staff with knowledge about the energy-efficient techniques or may only have operators in place part time to manage or oversee new technologies. Because operators generally are the advocates for energy-efficiency upgrades, the specialists believe it could be difficult to gain support for such investments without knowledgeable operators. Further, operators may be unaware of the amount of energy their facilities use because, in many municipalities, these bills are received and paid by other departments and operators may not have access to these data. Consequently, operators may be unaware of the potential for energy savings from upgrades. Moreover, many specialists told us that operators are often resistant to alter the practices that they have employed for years to move and treat water and may be reluctant to adopt new technologies or approaches, especially if the effectiveness of such changes has not yet been adequately proven. Some specialists also told us that drinking water and wastewater utilities do not coordinate as closely as they could with energy utilities to identify opportunities to optimize their operations and, thereby, lessen their energy demands.

Energy Usage Considerations Are Secondary to Complying with Water Quality Regulations

Considering the energy demands of treatment can be an afterthought to complying with water quality regulations for treatment plant operators, according to some specialists with whom we spoke. One drinking water utility operator told us that energy is considered to the extent possible when decisions are being made about altering treatment processes to meet regulatory requirements but that the safety of the water supply is his primary concern. For example, when the city of San Diego's Public Utilities Department was considering which disinfection technology to employ, it chose to use ozonation because it would provide more effective disinfection for the plant and also reduces disinfection by-products, even though it is a more energy-intensive technology than the current disinfection process. In addition, to ensure that minimum effluent discharge standards are met, water utility operators may over-treat wastewater by, for example, running aeration blowers at higher levels than necessary to meet regulatory requirements. In light of the potential for more stringent standards in the future, some specialists noted that regulators should consider the energy demands associated with these increased water quality standards.

Lack of Public Awareness about the Energy Demands of the Urban Water Lifecycle

Many specialists told us that many customers are not aware of and do not understand the energy demands of drinking water and wastewater services. While some customers may be aware of their total energy use, it may not be clear to them how much of that energy use is for heating water and other water-related uses. In addition, customers may not be aware that water conservation saves not only water but also energy. Some specialists told us that federal programs such as EPA's WaterSense and Energy Star and some state efforts, such as in California and New York, have begun to educate the public on the energy demands of the urban water lifecycle; however, additional efforts may be needed to increase awareness of the energy-water nexus for providing drinking water and wastewater to urban users.

Agency Comments

We provided a draft of this report to the Departments of Defense, Energy, and the Interior and EPA for review and comment. DOE and EPA provided technical comments that we incorporated into the final report as appropriate.

Sincerely yours,

Anu K. Mittal
Director, Natural Resources and Environment

Mark E. Gaffigan
Managing Director, Natural Resources and Environment

Appendix I. Objectives, Scope, and Methodology

Our objectives for this review were to describe what is known about (1) the energy needed for each stage of the urban water lifecycle, and (2) technologies and approaches that could lessen the energy needed for the urban water lifecycle, as well as any identified barriers that exist to their adoption. We focused our work on community drinking water systems and publicly owned wastewater facilities located in the United States. We also focused on residential customers and, to the extent possible, commercial, industrial, and institutional customers.

To address both of these objectives, we conducted a systematic review of studies and other documents that examine the energy required to extract, move, use, and treat water, including peer-reviewed scientific and industry periodicals, government-sponsored research, and reports from nongovernmental research organizations. In conducting this review, we searched databases such as ProQuest, EconLit, and BioDigest, and used an iterative process to identify additional studies, asking specialists to identify relevant studies and reviewing studies from article bibliographies. We reviewed studies that fit the following criteria for selection: (1) the research was of sufficient breadth and depth to provide observations or conclusions directly related to our objectives; (2) the research demonstrated the energy demands of water supply systems in the United States; (3) the studies typically were published between 2000 and 2010; and (4) the studies were determined to be methodologically sufficient. We examined key assumptions, methods, and relevant findings within the studies related to drinking water processes, customer end use, and wastewater processes. We believe we have included the key studies and have qualified our findings, where appropriate. However, it is possible that we may not have identified all of the studies with findings relevant to these two objectives.

We also selected a nonprobability sample of three cities to examine in greater depth to better understand regional and local differences related to urban water lifecycles: Memphis, Tennessee; San Diego, California; and Washington, D.C. We chose these cities as illustrative case studies based on criteria such as their type of water source; water availability; type of wastewater system; unique characteristics, such as potential for desalination; and economic factors, such as energy costs. The results from our visits to these cities cannot be generalized to all U.S. cities, but they provide valuable insights as illustrative case studies. For each of these case studies, we analyzed documentation from and conducted interviews with a wide range of specialists to gain the views of diverse organizations covering all stages of the urban water lifecycle. These groups included relevant drinking water and wastewater treatment facilities, and state and local agencies responsible for water or energy. We requested interviews with representatives from electrical utilities in each location. In San Diego and Washington, D.C., the utilities did not meet with us or told us they did not have relevant data. In Memphis, however, which has a combined water and energy utility, an energy official was present at our meeting with the utility, but the utility told us it does not track data on energy for water-related uses for some customer types. In addition, we conducted site visits to drinking water and wastewater treatment facilities in each of these locations to better understand the role that energy plays in their operation.

In addition to the specialists we interviewed as part of our illustrative case studies, we also interviewed a range of specialists whom we identified as having expertise related to the energy needs of all stages of the urban water lifecycle in general. We selected these specialists using an iterative process, soliciting additional names from each person we interviewed. From among those specialists identified, we interviewed those who could provide us with a broad range of perspectives on the energy needs of the urban water lifecycle.

We also interviewed specialists that we identified during our systematic review of studies who have analyzed (1) the energy needed in one or more stages of the water lifecycle at the national or local level or (2) techniques available to reduce the energy demands for water. These specialists represented a variety of organizations, including drinking water and wastewater treatment facilities; state and local government offices responsible for water or energy; officials from the EPA; researchers from some of the Department of Energy's national laboratories, such as Sandia National Laboratory; university researchers; water and energy industry representatives from groups such as the American Water Works Association and the Water Research Foundation; and relevant nongovernmental organizations, such as the Pacific Institute, a nonpartisan research institute that works to advance environmental protection, economic development, and social equity. The specialists also included individuals with knowledge of the energy demands for water in other states, including Arizona, Colorado, Florida, New York, and Wisconsin, to gain a better understanding of water and energy issues in other regions around the United States. We also interviewed other federal agency officials and analyzed data and information from federal agencies that have responsibilities related to the energy needs of the urban water lifecycle—the Department of Defense's U.S. Army Corps of Engineers, the Department of Energy, the Department of the Interior's U.S. Geological Survey and Bureau of Reclamation, the Environmental Protection Agency, and the National Science Foundation.

To analyze information gathered through the interviews with specialists and the scientific studies, research, and other key documents reviewed, we conducted content analyses. Specifically, to conduct the content analysis of information gathered through interviews with specialists, we reviewed each interview, selected relevant statements, and identified and labeled these statements using a coding system that identified the topic area. Once relevant statements from the interviews were extracted and coded, we used the coded data to develop key themes. An independent reviewer then verified that the codes were accurately applied to the statements and the key themes were correctly developed. During the course of our review, we conducted over 60 interviews with over 100 specialists. For the purposes of our interview analysis, each interview represents the views of one specialist even if more than one specialist was present at the interview. We used the following categories to quantify responses of experts and officials: "some" refers to responses from 2 to 5 specialists, "several" refers to responses from 6 to 10 specialists, and "many" refers to responses from 11 or more specialists. We used a similar coding scheme to identify key themes resulting from our analysis of the scientific studies, research, and other key relevant documentation.

We performed our work from January 2010 to January 2011 in accordance with all sections of GAO's Quality Assurance Framework that are relevant to our objectives. The framework requires that we plan and perform the engagement to obtain sufficient and appropriate evidence to meet our stated objectives and to discuss any limitations in our work. We believe that the information and data obtained, and the analysis conducted, provide a reasonable basis for any findings and conclusions in this product.

End Notes

[1] A metropolitan area, as defined by the Office of Management and Budget, consists of one or more counties containing at least one urbanized area of 50,000 or more people.

[2] For the purposes of this report, "urban" refers to areas of the country that are connected to community water systems and that receive wastewater services from municipal wastewater treatment facilities. It does not include agricultural water use or customers who self-supply water or rely on septic systems for waste disposal.

[3] 42 U.S.C. §§ 300f-300j-26 (2006).

[4] 33 U.S.C. §§ 1251-1387 (2006).

[5] We requested interviews with representatives from the primary electrical utilities in each location. In San Diego and Washington, D.C., the utilities did not provide representatives to meet with us or told us they did not have relevant data. In Memphis, however, which has a combined water and energy utility, an energy official was present at our meeting with the utility, but the utility told us it does not track data on energy for water-related uses for some customer types.

[6] A community water system is one that provides water for human consumption through pipes or other constructed conveyances to at least 15 service connections, or regularly serves at least 25 people year-round. Customers not part of a community water system can receive water from other types of public water systems or from private wells.

[7] Wastewater may be treated through other systems, such as septic systems, which serve approximately 25 percent of U.S. households.

[8] EPA may promulgate a treatment technique in lieu of a maximum contaminant level, if the EPA Administrator makes a finding that it is not economically or technologically feasible to ascertain the level of a given contaminant in drinking water.

[9] 33 U.S.C. §1342(q) (2006) (implementing the Combined Sewer Overflow Control Policy signed by the Administrator on April 11, 1994; see 59 Fed. Reg. 18688 (Apr. 19, 1994)). The policy defines a combined sewer system as a wastewater collection system owned by a state or municipality that conveys sanitary wastewaters and stormwater through a single-pipe system to a publicly owned treatment works treatment plant.

[10] According to some specialists we spoke with, topography and distance can also affect the energy needed to pump and move wastewaster. However, the specialists noted that wastewater systems were often designed to rely on gravity, and wastewater treatment plants were located close to the receiving waters.

[11] Advanced primary treatment includes enhanced removal of solids and organic matter from wastewater, which is typically accomplished by chemical addition or filtration.

[12] A trickling filter is a bed of media, typically rocks or plastic, through which the wastewater passes. A lagoon system uses a scientifically constructed pond that allows sunlight, algae, bacteria, and oxygen to interact and treat the wastewater.

[13] Treatment plants may also install ultraviolet light disinfection technologies for reasons unrelated to water quality, such as concerns about plant safety and security when storing large quantities of chemicals on-site.

[14] Environmental Protection Agency, *Evaluation of Energy Conservation Measures for Wastewater Treatment Facilities* (Washington, D.C., 2010).

[15] WaterSense is an EPA-sponsored partnership program that seeks to protect the future of our nation's water supply by promoting water efficiency and enhancing the market for water-efficient products, programs, and practices.

[16] Biogas is a mixture of gases including methane and carbon dioxide produced during the digestion of organic solids that result from the wastewater treatment process.

[17] American Recovery and Reinvestment Act of 2009, Pub. L. No. 111-5, 123 Stat. 115, 169; Interior Department and Further Continuing Appropriations, Fiscal Year 2010, 123 Stat. 2904, 2935.

In: Further Explorations in the Energy-Water Nexus ISBN 978-1-62081-318-8
Editors: D. A. Hughes and B. F. Smith

Chapter 4

A REVIEW OF OPERATIONAL WATER CONSUMPTION AND WITHDRAWAL FACTORS FOR ELECTRICITY GENERATING TECHNOLOGIES*

Jordan Macknick, Robin Newmark, Garvin Heath and KC Hallett

ACKNOWLEDGMENTS

This work was funded by the U.S. Department of Energy's (DOE's) Office of Energy Efficiency and Renewable Energy (EERE) and Office of Policy and International Affairs (PI). The authors wish to thank Allan Hoffman and Diana Bauer for their support of this work. We are also indebted to the following individuals for their thoughtful comments, input, and review of the document in its various stages: Kristen Averyt, Western Water Assessment (WWA) and the University of Colorado; Stacy Tellinghuisen, Western Resource Advocates; Timothy Diehl, U.S. Geological Survey; and Lynn Billman, Elaine Hale, Margaret Mann, Walter Short, and Daniel Steinberg, NREL. In addition, we would like to thank the participants in the

---

* This is an edited, reformatted and augmented version of the U.S. Department of Energy, Office of Energy Efficiency & Renewable Energy, dated March 2011.

Water for Energy Workshop who provided valuable input, particularly Christina Alvord and Brad Udall, WWA and University of Colorado; Mike Hightower and Vince Tidwell, Sandia National Laboratories; Curt Brown, U.S. Bureau of Reclamation; Margot Gerritsen, Stanford University; Eric Fournier, UC Santa Barbara; Alex Schroeder, Western Governors' Association; Ashlynn Stillwell, University of Texas Austin; Steve Clemmer and John Rogers, Union of Concerned Scientists; Andrew Wolfsberg, Los Alamos National Laboratory; and Larry Flowers, NREL. We also wish to thank Mary Lukkonen of NREL for her editorial support.

EXECUTIVE SUMMARY

This report provides estimates of operational water withdrawal and water consumption factors for electricity generating technologies in the United States. Estimates of water factors were collected from published primary literature and were not modified except for unit conversions. The presented water factors may be useful in modeling and policy analyses where reliable power plant level data are not available. Major findings of the report include:

- The power sector withdraws more water than any other sector in the United States and is heavily dependent on available water resources. Changes in water resources may impact the reliability of power generation.
- Water withdrawal and consumption factors vary greatly across and within fuel technologies. Water factors show greater agreement when organized according to cooling technologies as opposed to fuel technologies. Once-through cooling technologies withdraw 10 to 100 times more water per unit of electric generation than recirculating cooling technologies; recirculating cooling technologies consume at least twice as much water as once-through cooling technologies.
- A transition to a less carbon-intensive electricity sector could result in either an increase or decrease in water use, depending on the choice of technologies and cooling systems employed. Concentrating solar power (CSP) technologies and coal facilities with carbon capture and sequestration (CCS) capabilities have the highest water consumption values when using a recirculating cooling system. Non-thermal renewables, such as photovoltaics (PV) and wind, have the lowest water consumption factors.

- Federal datasets on water use in power plants have numerous gaps and methodological inconsistencies. Federal agencies are currently coordinating to improve these data. Water use factors discussed here are good proxies for use in modeling and policy analyses, at least until power plant level data improve.
- Impacts of the power sector on freshwater availability can be reduced by utilizing dry cooling or by using non-freshwater sources for cooling. However, these alternatives are limited by locally available resources and may have cost and performance penalties.

Improved power plant data and further studies into the water requirements of energy technologies in different climatic regions would facilitate greater resolution in analyses of water impacts of future energy and economic scenarios. This report provides the foundation for conducting water use impact assessments of the power sector while also identifying gaps in data that could guide future research.

1. INTRODUCTION

Thermoelectric power use has a significant impact on water resources and the power sector is highly dependent on these water resources; the United States Geological Survey (USGS) estimated on a national level that 41% of all freshwater withdrawals in the United States in 2005 were for thermoelectric power operations, primarily for cooling needs [1]. The power sector is thus highly vulnerable to changes in water resources, especially those that may result from potential climatic changes [2-5]. Increasingly, state agencies in California and New York have taken policy actions to address the impacts of power plants' water use and the environmental impacts of their cooling systems [6, 7]. Furthermore, the 2007 drought in the Southeast exposed many thermal generators, including Brown's Ferry nuclear plant, to water-related shut downs and curtailments due to unlawfully high discharge temperature and shallow or exposed cooling water inlet locations [8]. Effective integrated energy and water policy planning will require identifying the individual and cumulative impacts that power plant configurations have on water resources and the vulnerabilities of specific power plants to changes in water resources. Various studies have attempted to consolidate published estimates of water use impacts of electricity generating technologies, resulting in a wide range of technologies and values based on different primary sources of literature [9-14].

The goal of this work is to consolidate the various primary literature estimates of water use during the generation of electricity by conventional and renewable electricity generating technologies in the United States to more completely convey the variability and uncertainty associated with water use in electricity generating technologies. Individual water use factors, reported in terms of the volume of water used per unit of electrical output (gallons per megawatt-hour), are technology and cooling system specific. These water use factors can be incorporated into energy-economic models to estimate generation-related water use under different projected electricity portfolio scenarios.

2. Scope and Methods

We evaluate two aspects of water usage: withdrawal and consumption. According to the USGS, "withdrawal" is defined as the amount of water removed from the ground or diverted from a water source for use, while "consumption" refers to the amount of water that is evaporated, transpired, incorporated into products or crops, or otherwise removed from the immediate water environment [1]. Both water withdrawal and consumption values are important indicators for water managers determining power plant impacts and vulnerabilities associated with water resources.

We consider water withdrawals and consumption for the operational phase only. Operational water use in this study includes cleaning, cooling, and other process-related needs that occur during electricity generation, such as flue gas desulfurization (FGD) in coal facilities. For the vast majority of power generation technologies, most of the water used in the life cycle of the plant occurs during the operational phase, with the exception of non-thermal renewable energy technologies that do not require cooling systems [9]. In addition, compared to the operational phase, data for the water requirements of other phases (such as the fuel cycle) are scarce, are subject to greater definitional boundary differences, and have more site-specific differences. Also, although the location of the plant is permanent, the locations of the manufacturing or fuel sources are not permanent. Given this and the continuous local impacts of power plant water use on water resources during the operational phase, we limit this study to a detailed review of only the operational water requirements of electricity generating technologies.

The energy technologies addressed here consist of configurations of concentrating solar power (CSP), solar photovoltaic (PV), wind, biopower, geothermal, hydroelectric, nuclear, natural gas, and coal technologies. Cooling system technologies considered include wet recirculating technologies (evaporative cooling towers), once-through cooling systems (open loop cooling), air-cooled condensing (dry cooling), hybrid wet and dry cooling systems (hybrid cooling), and pond cooling systems.

Electricity generating technologies use water for different processes, depending on their configuration. Thermal electricity technologies (e.g., CSP, biopower, coal, nuclear, and natural gas technologies) generally require water as the working fluid (and as the cooling medium to condense steam) as part of the Rankine cycle, the thermodynamic process that drives the steam engine [15]. CSP facilities use water for steam cycle processes, for cleaning mirrors or heliostats, and for cooling if a cooling tower is used. PV systems require occasional panel washing. Wind systems require very little water, if any, for cleaning. Biopower facilities use water for cooling and for steam cycle processing. Upstream water needs for growing energy crops are not included in this analysis but can be quite substantial (approximately 100 times greater than operational cooling system needs) and can vary greatly depending on region, crop, and production methods [16-18]. Geothermal technology configurations (e.g., dry steam, binary, and flash) can differ greatly in their use of water due to differences in reinjection techniques as well as vapor temperature and mass [19]. Enhanced Geothermal Systems (EGS) operate similar to geothermal binary technologies yet also require some additional water for hydraulic stimulation [19]. Water used in geothermal technologies may come from geothermal fluids, with little to no impact on local freshwater sources [20]. Over time, however, some geothermal plant efficiencies may decline and may require outside fresh or brackish water sources, and some technologies may lower local water tables [19, 21]. Hydroelectric facilities using reservoirs have evaporative losses resulting from the dammed water [22, 23]. Nuclear, natural gas, and coal facilities use water for cooling and for steam cycle processes. Coal facilities may also use water for FGD. Fossil technologies employing carbon capture and storage capabilities will require additional process water requirements [12].

Estimates of water consumption and withdrawal are displayed irrespective of geographic location, as many published data do not specify the location or climatic conditions of the plant. The location of a plant, and its corresponding climatic conditions, can affect its overall efficiency and thus its water use rate [24-27]. Similar fossil plants utilizing cooling towers may have water consumption and withdrawal factors that differ by more than 16%, depending on the location in the United States [28]. Similarly, water consumption factors of CSP plants utilizing cooling towers may differ by as much as 20% [15]. Inter-annual variations in water intensity are also not considered for this review. Withdrawal and consumption factors are often reported in terms of annual averages, yet water intensity of facilities may change by as much as 16% as a result of diurnal and seasonal variations in temperatures, wind speeds, and humidity levels [28]. Other factors that may influence water use intensities of power plants that are not considered here include the age of the plant, the thermal efficiency of the plant, the age of the cooling system, and the water source [26, 27].

Certain aggregations of fuel technology types and cooling system types were made to facilitate analyses. Nuclear technologies include pressurized water reactors and boiling water reactors. Coal technologies make no distinction between wet, dry, and no FGD. For recirculating cooling technologies, no distinction is made between natural draft and mechanical draft cooling tower systems. All pond-cooled systems are treated identically. Pond-cooled systems can be operated in manners that resemble both recirculating systems and once-through systems as well as in hybrids of these technologies [29]. Different configurations and operating practices of pond-cooled systems can lead to widely different reported water withdrawal and consumption values. No distinction is made between water types, which may include freshwater (surface and groundwater), saline water, or municipal waste water. In 2005, 71% of thermoelectric water withdrawals were from freshwater sources [1]. Saline withdrawals are primarily concentrated in California, Florida, and the coastal Northeast, with the rest of the country relying on freshwater.

Data sources include published academic literature, state and federal government agency reports, non-governmental organizations' reports, and industry submissions to government agencies for permitting procedures. Estimates of national average water use intensity for particular technologies, estimates of existing plant operational water use, and estimates derived from laboratory experiments were considered equally. Certain sources report ranges of water consumption and withdrawal factors in place of specific values. If traceable individual case studies form the basis for the range given, the individual values are included as independent estimates within the set of estimates that are statistically analyzed. If a range is given and the underlying data points are not given, then the midpoint of that range is used for calculating an average value, and the high and low extremes are used for determining extreme ranges. This method of addressing ranges may lead to a bias toward data sources reporting explicit cases and may also underestimate actual water use at facilities, as it was observed that the midpoint of the range of extremes are in general less than values reported from individual facilities. This review did not alter (except for unit conversion) or audit for accuracy the estimates of water use published. Because estimates are used as published, considerable methodological inconsistency is inherent, limiting comparability. We report minimum, maximum, and median values for fuel technology and cooling system combinations in tables and additionally show 25th and 75th percentile data in figures. Due to the wide range of values reported from a small number of sources, median values may differ significantly from mean values. Upon request, raw data are available from the authors.

3. Data Availability and Gaps

Although the power sector is the largest user of water in the nation, national statistics on the consumption and withdrawal rates of individual power plants are characterized by inconsistencies and scarcity [30]. Power sector water use data on a national level are collected by two federal agencies, the USGS and the U.S. Department of Energy's Energy Information Administration (EIA). The USGS reports water withdrawals for thermoelectric power production by county and sector every five years; water consumption values for thermoelectric power production were last reported for 1995 [31]. These data are collected by state agencies that do not always utilize the same methods or definitions in determining water withdrawals [1].

EIA provides official energy statistics on an annual basis, and EIA Form 923 reports, among other data, annual water withdrawal, discharge, and consumption rates in Schedule 8D, providing similar definitions of withdrawal and consumption as the USGS [29]. However, data are not entirely comprehensive and have omitted nuclear facilities and some natural gas combined cycle technologies [32]. Additionally, the quality of data is also of concern with power plants reporting data; many of the power plants report water withdrawal and consumption values that are far below or above detailed engineering studies of water use in power plants considered in this review. The National Energy Technology Laboratory compiled water use data in their *2007 Coal Power Plant DataBase* [33]. However, this database is limited by the data availability and quality of EIA datasets. No similar public database has been developed for natural gas or nuclear generating facilities.

Detailed engineering studies and more general assessments of water use at individual thermoelectric power plants are uneven in their treatment of fuel technologies and cooling systems. For example, water consumption data for coal, natural gas, nuclear, and parabolic trough CSP facilities using a wet recirculating cooling system are relatively abundant. Fewer studies are available addressing water withdrawals for all technologies or water consumption for once-through, pond, and dry-cooling systems. Very little data exist for dedicated biomass, geothermal, and power tower CSP facilities.

Additionally, boundary conditions of water use studies are not always clear or consistent; some sources only report aggregated operational water usage, whereas other reports include water use by individual processes. However, the particular processes included in disaggregated studies may not be equivalent; the inclusion of FGD water requirements in coal facilities is one example where its explicit or implicit consideration is inconsistent across datasets. Geothermal facilities add an additional layer of complexity, as often cooling processes can make use of geothermal fluids rather than freshwater; some sources exclude geothermal fluids from calculations whereas others include geothermal fluids. Estimates of evaporation from hydropower reservoirs are complicated by the multiple uses of reservoirs (e.g., water supply, recreation, and flood control) and the different methods of allocating evaporation to electricity production [22, 23]. Hydropower estimates are reported according to the allocation methods utilized in the published reports, which allocate all reservoir evaporation to power production.

4. Results: Water Consumption and Withdrawal Factors

The cooling system employed is often a greater determinant of water usage than the particular technology generating electricity, both in terms of water consumption (Figures 1 and 2) and water withdrawal (Figures 3 and 4). Once-through cooling technologies withdraw 10 to 100 times more water per unit of electric generation than cooling tower technologies, yet cooling tower technologies consume at least twice as much water as once-through cooling technologies. Water consumption for dry cooling at CSP, biopower, and natural gas combined cycle plants is an order of magnitude less than for recirculating cooling at each of those types of plants.

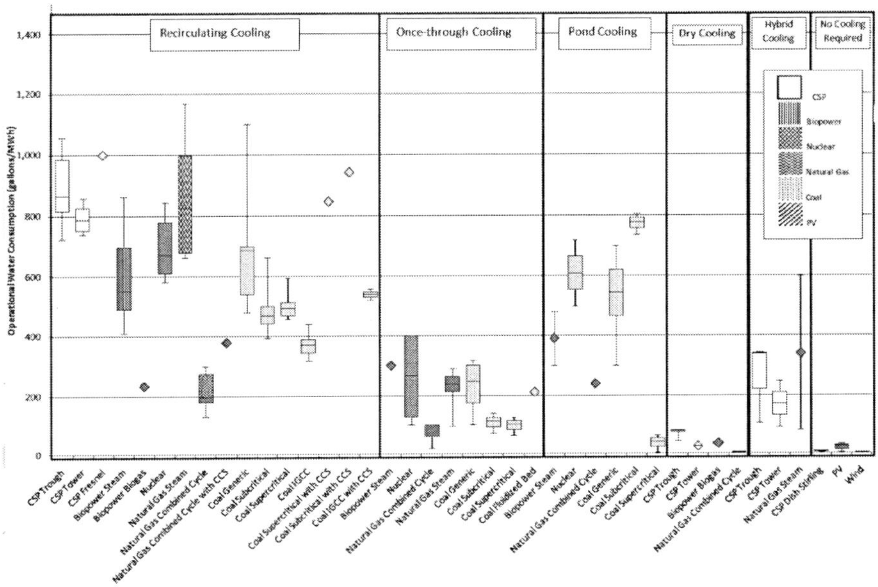

IGCC: Integrated gasification combined cycle. CCS: Carbon capture and sequestration. CSP: Concentrating solar power. Whisker ends represent maxima and minima. Upper and lower ends of boxes represent 75th and 25th percentile, respectively. Horizontal lines in boxes represent medians.

Figure 1. Operational water consumption factors for electricity generating technologies.

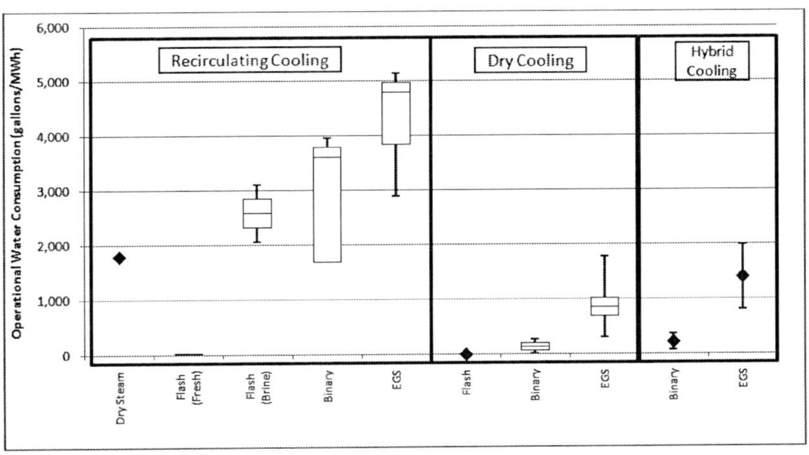

EGS: Enhanced geothermal systems. Whisker ends represent maxima and minima. Upper and lower ends of boxes represent 75th and 25th percentile, respectively. Horizontal lines in boxes represent medians.

Figure 2. Operational water consumption factors for geothermal technologies.

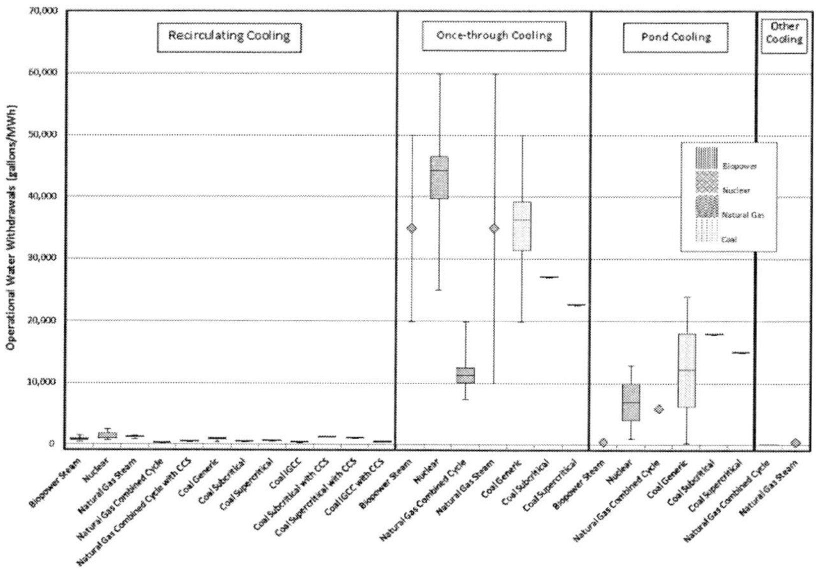

IGCC: Integrated gasification combined cycle. CCS: carbon capture and storage. Whisker ends represent maxima and minima. Upper and lower ends of boxes represent 75th and 25th percentile, respectively. Horizontal lines in boxes represent medians. Recirculating cooling withdrawal values are also shown in Figure 4.

Figure 3. Operational water withdrawals for electricity generating technologies.

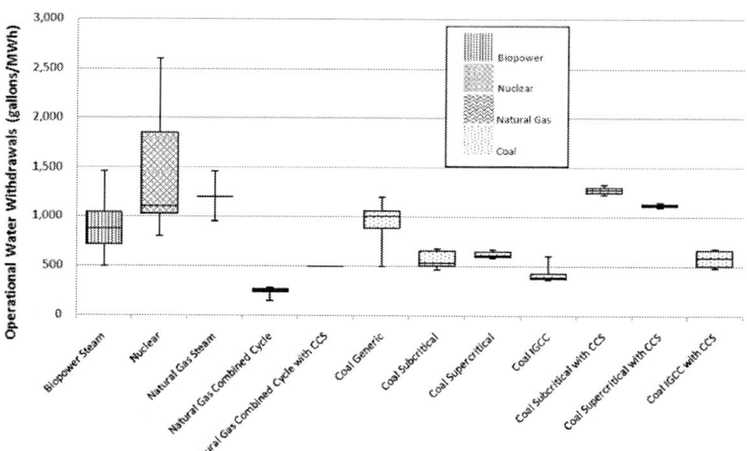

IGCC: Integrated gasification combined cycle. CCS: carbon capture and storage. Whisker ends represent maxima and minima. Upper and lower ends of boxes represent 75th and 25th percentile, respectively. Horizontal lines in boxes represent medians.

Figure 4. Operational water withdrawal factors for recirculating cooling technologies.

Table 1. Water Consumption Factors for Renewable Technologies (gal/MWh)

Fuel Type	Cooling	Technology	Median	Min	Max	*n*	Sources
PV	N/A	Utility Scale PV	26	0	33	3	[10, 34, 35]
Wind	N/A	Wind-Turbine	0	0	1	2	[11, 36]
CSP	Tower	Trough	865	725	1,057	17	[10, 34, 37-46]
		Power Tower	786	740	860	4	[34, 39-41]
		Fresnel	1,000	1,000	1,000	1	[47]
	Dry	Trough	78	43	79	10	[38, 42-44]
		Power Tower	26	26	26	1	[48]
	Hybrid	Trough	338	105	345	3	[42, 47]
		Power Tower	170	90	250	2	[47]
	N/A	Stirling	5	4	6	2	[34, 49]
Biopower	Tower	Steam	553	480	965	4	[49-51]
		Biogas	235	235	235	1	[52]
	Once-through	Steam	300	300	300	1	[50]
	Pond	Steam	390	300	480	1	[50]
	Dry	Biogas	35	35	35	1	[51]
Geothermal[1]	Tower	Dry Steam	1,796	1,796	1,796	1	[10]
		Flash (freshwater)	10	5	19	3	[19, 20, 49]
		Flash (geo-thermal fluid)	2,583	2,067	3,100	2	[53]
		Binary	3,600	1,700	3,963	3	[10, 54, 55]
		EGS	4,784	2,885	5,147	4	[10, 51, 54, 55]
	Dry	Flash	0	0	0	1	[51]
		Binary	135	0	270	2	[19, 51]
		EGS	850	300	1,778	2	[19, 51]
	Hybrid	Binary	221	74	368	1	[56]
		EGS	1,406	813	1,999	2	[51, 56]
Hydropower	N/A	Aggregated in-stream and reservoir	4,491	1,425	18,000	3	[22, 23]

[1] Most geothermal facilities can use geothermal fluids or freshwater for cooling.

Water consumption factors for renewable (Table 1) and non-renewable (Table 2) electricity generating technologies vary substantially within and across technology categories. The highest water consumption factors for all technologies result from the use of evaporative cooling towers. With the exception of hydropower, pulverized coal with carbon capture and CSP technologies utilizing a cooling tower represent the upper bound of water consumption, at approximately 1,000 gal/MWh of electricity production. The lowest operational water consumption factors result from wind energy, PV, and CSP Stirling solar technologies and natural gas combined cycle facilities that employ dry cooling technologies. Water withdrawal factors for electricity generating technologies show a similar variability within and across technology categories (Table 3). The highest water withdrawal values result from nuclear technologies, whereas the smallest withdrawal values are for non-thermal renewable technologies. Consistent with literature, withdrawal factors for CSP, wind, geothermal, and PV systems are assumed to be equivalent to consumption factors.

Table 2. Water Consumption Factors for Non-renewable Technologies (gal/MWh)

Fuel Type	Cooling	Technology	Median	Min	Max	n	Sources
Nuclear	Tower	Generic	672	581	845	6	[10, 14, 27, 50, 57]
	Once-through	Generic	269	100	400	4	[27, 50, 57, 58]
	Pond	Generic	610	560	720	2	[27, 50]
Natural Gas	Tower	Combined Cycle	198	130	300	5	[13, 34, 50, 57, 59]
		Steam	826	662	1,170	4	[10, 14, 49, 60]
		Combined Cycle with CCS	378	378	378	1	[59]
	Once-through	Combined Cycle	100	20	100	3	[50, 57, 60]
		Steam	240	95	291	2	[10, 49]
	Pond	Combined Cycle	240	240	240	1	[57]
	Dry	Combined Cycle	2	0	4	2	[50, 57]
	Inlet	Steam	340	80	600	1	[49]
Coal	Tower	Generic	687	480	1,100	5	[10, 14, 27, 50, 58]
		Subcritical	471	394	664	6	[13, 57, 59, 61]
		Supercritical	493	458	594	6	[13, 57, 59, 61]
		IGCC	372	318	439	7	[13, 59]
		Subcritical with CCS	942	942	942	1	[59]
		Supercritical with CCS	846	846	846	1	[59]
		IGCC with CCS	540	522	558	3	[59]
	Once-through	Generic	250	100	317	4	[10, 27, 50, 58]
		Subcritical	113	71	138	3	[57]
		Supercritical	103	64	124	3	[57]
	Pond	Generic	545	300	700	2	[27, 50]
		Subcritical	779	737	804	3	[57]
		Supercritical	42	4	64	3	[57]

Table 3. Water Withdrawal Factors for Electricity Generating Technologies (gal/MWh)

Fuel Type	Cooling	Technology	Median	Min	Max	n	Sources
Nuclear	Tower	Generic	1,101	800	2,600	3	[27, 50, 57]
	Once-through	Generic	44,350	25,000	60,000	4	[27, 50, 57, 58]
	Pond	Generic	7,050	500	13,000	2	[27, 50]
Natural Gas	Tower	Combined Cycle	253	150	283	6	[12, 13, 50, 57, 59]
		Steam	1,203	950	1,460	2	[49, 60]
		Combined Cycle with CCS	496	487	506	2	[12, 59]
	Once-through	Combined Cycle	11,380	7,500	20,000	2	[50, 57]
		Steam	35,000	10,000	60,000	1	[49]
	Pond	Combined Cycle	5,950	5,950	5,950	1	[57]
	Dry	Combined Cycle	2	0	4	2	[50, 57]
	Inlet	Steam	425	100	750	1	[49]
Coal	Tower	Generic	1,005	500	1,200	4	[27, 35, 50, 58]
		Subcritical	531	463	678	7	[12, 13, 57, 59, 61]
		Supercritical	609	582	669	7	[12, 13, 57, 59, 61]
		IGCC	390	358	605	11	[12, 13, 35, 59]
		Subcritical with CCS	1,277	1,224	1,329	2	[12, 59]
		Supercritical with CCS	1,123	1,098	1,148	2	[12, 59]
		IGCC with CCS	586	479	678	6	[12, 59]
	Once-through	Generic	36,350	20,000	50,000	4	[11, 27, 50, 58]
		Subcritical	27,088	27,046	27,113	3	[57]
		Supercritical	22,590	22,551	22,611	3	[57]
	Pond	Generic	12,225	300	24,000	2	[27, 50]
		Subcritical	17,914	17,859	17,927	3	[57]
		Supercritical	15,046	14,996	15,057	3	[57]
Biopower	Tower	Steam	878	500	1,460	2	[49]
	Once-through	Steam	35,000	20,000	50,000	1	[50]
	Pond	Steam	450	300	600	1	[50]

5. Discussion

Despite methodological differences in data, general trends can be observed and broad conclusions can be drawn from the breadth of data collected. A transition to a less carbon-intensive electricity sector could result in either an increase or decrease in water consumption per unit of electricity generated, depending on the choice of technologies and cooling systems employed. Non-thermal renewable technologies, such as wind and PV systems, consume minimal amounts of water per unit of generation. However, the highest water consumption factors considered in this study, excluding geothermal and hydroelectric facilities, which can have high water intensities but also have important caveats, are low-carbon emitting technologies that utilize cooling towers: pulverized coal with carbon capture technologies and CSP systems. Decisions affecting the power sector's impact on the climate may need to include water considerations to avoid negative unintended environmental consequences on water resources. This can be addressed by integrated energy and water policy planning, as the availability of water in certain jurisdictions may limit the penetration of these technologies and cooling system configurations.

Freshwater use impacts can be reduced by utilizing dry cooling or by using non-freshwater sources as a cooling medium. Initial work suggests that the performance penalty for CSP facilities switching from wet cooling to dry cooling results in an annual reduction in output of 2%–5% and an increase in the levelized cost of producing energy of 3%–8%, depending on local climatic conditions [15]. Using national averages, the annual performance penalty for switching from wet cooling to dry cooling for nuclear plants is 6.8%, combined cycle plants 1.7%, and other fossil plants (including coal and natural gas steam plants) 6.9% [62]. Further efforts are needed to evaluate performance and cost penalties associated with utilizing dry or hybrid cooling systems for fossil fuel facilities using carbon capture technologies. Utilizing reclaimed water, such as municipal wastewater, is another approach that could lessen the impact of the power sector on freshwater resources and wastewater treatment facilities. The legal and physical availability of municipal wastewater, especially in rural areas, may be a limiting factor to its widespread usage, and the cost and performance penalties of utilizing such sources must be investigated further [63].

The choice of cooling system may play an important role in the development of our future electricity mix. Differences between cooling systems can have substantial environmental impacts on local water resources

[64-66]. Employing wet cooling technologies (i.e., once-through and cooling tower technologies) imposes an inherent tradeoff between relatively high water consumption and relatively high water withdrawals, which has important implications for regional cooling system policies and regulations. A reduction in withdrawals (but a corresponding increase in consumption) may benefit a watershed that has an abundance of water but may lead to concerns in an area that is already lacking water. A shift away from, for example, once-through cooling systems in coastal areas that withdraw saline water, to inland recirculating systems such as cooling towers that primarily consume freshwater, will impact watersheds and water availability differently depending on local conditions. The use of alternative cooling technologies may serve as an energy security benefit for utilities and communities, given uncertainties in future scenarios of water availability and expected vulnerabilities for power plants [4, 5]. Reduced levels in bodies of water, or substantial increases in the temperature of these bodies of water, may require thermal power plants to run at lower capacities or to shut down completely, as was seen in France in 2003 [67]. Utilizing dry cooling or non-freshwater sources avoids some of the risks associated with these drought and climate change scenarios.

Accurate estimates of water use in individual power plants, and the effect of this water use on a regional scale, will be elusive until more studies are conducted for the variety of technologies and cooling systems currently in operation along with those expected to be developed and deployed. Furthermore, calibration of these values on national and regional scales will remain challenging until methods for collecting and evaluating data by federal agencies has improved. Nonetheless, certain conclusions regarding the overall impact power plants have on water resources can be drawn on regional levels from existing water use data.

Further studies with consistent boundary conditions and methods are necessary to develop water consumption and withdrawal estimates for certain technologies and cooling systems to fully understand reasons for variations in data that are not attributable to climatic factors or technology vintages. To better understand how cooling system and technology system decisions will be made in the future, analyses using energy-economic models will require improved data on water availability and regional water use factors. Existing data collected from federal agencies are currently inconsistent and incomplete [30]. However, in 2009, the U.S. Government Accountability Office released a report calling for improvements in federal agency water data collection in power plants; EIA is currently working with the USGS and other federal agencies to improve the scope and quality of its data collection [30]. Such efforts should improve the availability of power plant specific data and the ability to calibrate model estimates.

SUMMARY

We reviewed primary literature for data on water withdrawal and consumption factors for electricity generation in the United States and have consolidated them in this study. These detailed water consumption and withdrawal factors can be utilized in energy-economic and transmission planning models to better understand the regional and national impacts on water resources for various electricity future scenarios and can inform policy analysis at a national and local level. Improved power plant data gathered on a regional level and further studies into the water requirements of existing and emerging technologies (such as carbon capture technologies) are necessary to assess the water impacts of a developing decarbonizing economy in more detail.

REFERENCES

[1] Kenny, J.F.; Barber, N.L.; Hutson, S.S.; Linsey, K.S.; Lovelace, J.K.; Maupin, M.A. *Estimated Use of Water in the United States in 2005.* U.S. Geological Survey Circular 1344. Reston, VA: USGS, 2009; p. 52.

[2] Vörösmarty, C.; Green, P.; Salisbury, J.; Lammers, R. *"Global Water Resources: Vulnerability from Climate Change and Population Growth." Sci.*; Vol. 289 (5477), 2000.

[3] Bates, B.C.; Kundzewicz, Z.W.; Wu, S.; Palutikof, J.P. *Climate Change and Water. Technical Paper of the Intergovernmental Panel on Climate Change*. Geneva, Switzerland: IPCC, 2008.

[4] Dai, A. "Drought Under Global Warming: A Review." *Wiley Interdiscip. Rev.: Clim. Change*; Vol. 2, 2010; pp. 45–65.

[5] National Energy Technology Laboratory (NETL). *Water Vulnerabilities for Existing Coal-fired Power Plants.* DOE/NETL-2010/1429. Pittsburgh, PA: National Energy Technology Laboratory, 2010.

[6] California State Lands Commission. *Resolution by the California State Lands Commission Regarding Once-through Cooling in California Power Plants*. Proposed April 13, 2006.

[7] New York State Department of Environmental Conservation. *Best Technology Available (BTA) for Cooling Water Intake Structures*. Proposed March 4, 2010.

[8] NETL. *Impact of Drought on U.S. Steam Electric Power Plant Cooling Water Intakes and Related Water Resource Management Issues*. DOE/NETL-2009/1364. Pittsburgh, PA: National Energy Technology Laboratory, 2009.

[9] Fthenakis, V.; Kim, H.C. "Life-cycle Uses of Water in U.S. Electricity Generation." *Renew. Sustain. Energy Rev.;* Vol. 14, 2010; pp. 2039–2048.

[10] Gleick, P. *Water in Crisis: A Guide to the World's Fresh Water Resources*. New York: Oxford University Press, 1993.

[11] Inhaber, H. "Water Use in Renewable and Conventional Electricity Production." *Energy Sources, Part A;* Vol. 26, 2004; pp. 309–322.

[12] NETL. *Cost and Performance Baseline for Fossil Energy Plants-Volume 1: Bituminous Coal and Natural Gas to Electricity Final Report*. DOE/NETL-2007/1281. Pittsburgh, PA: National Energy Technology Laboratory, 2007.

[13] NETL. *Power Plant Water Usage and Loss Study. 2007 Update*. Pittsburgh, PA: National Energy Technology Laboratory, 2007.

[14] Western Resource Advocates (WRA). *A Sustainable Path: Meeting Nevada's Water and Energy Demands*. Boulder, CO: Western Resource Advocates, 2008.

[15] Turchi, C.; Wagner, M.; Kutscher, C. *Water Use in Parabolic Trough Power Plants: Summary Results from WorleyParsons' Analyses*. NREL/TP-5500-49468. Golden, CO: National Renewable Energy Laboratory, 2010.

[16] Stone, K.C.; Hunt, P.G.; Cantrell, K.B.; Ro, K.S. "The Potential Impacts of Biomass Feedstock Production on Water Resource Availability." *Bioresour.Technol.;* Vol. 101, 2010; pp. 2014–2025.

[17] Berndes, G. "Bioenergy and Water—The Implications of Large-scale Bioenergy Production for Water Use and Supply." *Global Environ. Change;* Vol. 12, 2002; pp. 253–271.

[18] Berndes, G. "Future Biomass Energy Supply: The Consumptive Water Use Perspective." *Int. J. Water Resour. Dev;* Vol. 24, 2008; pp. 235–245.

[19] Clark, C.; Harto, C.; Sullivan, J.; Wang, M. *Water Use in the Development and Operation of Geothermal Power Plants.* ANL/EVS/R-10/5. Argonne, IL: Argonne National Laboratory, 2011.

[20] Kagel, A.; Bates, D.; Gawell, K. *A Guide to Geothermal Energy and the Environment.* Washington, DC: Geothermal Energy Association, 2007.

[21] Bradbury, D. "Californian Groups Clash Over Geothermal Water Use." *Bus. Green*; 2009.

[22] Gleick, P. "Environmental Consequences of Hydroelectric Development: The Role of Facility Size and Type." *Energy*; Vol. 17 (8), 1992; pp. 735–747.

[23] Torcellini, P.; Long, N.; Judkoff, R. *Consumptive Water Use for U.S. Power Production.* NREL/TP-550-33905. Golden, CO: National Renewable Energy Laboratory, 2003.

[24] Miller, B.A.; Alavian, V.; Bender, M.D.; Benton, D.J.; Ostrowski Jr., P.; Parsly, J.A.; Shiao, M.C. "Integrated Assessment of Temperature Change Impacts on the TVA Reservoir and Power Supply Systems." *Hydraulic Engineering: Saving a Threatened Resource—In Search of Solutions: Proceedings of the Hydraulic Engineering Sessions at Water Forum '92;* 1992; pp. 563–568.

[25] Giusti, E.; Meyer, E. *Water Consumption by Nuclear Power Plants and Some Hydrological Implications.* U.S. Geological Survey Circular 745. Reston, VA: USGS, 1977.

[26] Yang, X.; Dziegielewski, B. "Water Use by Thermoelectric Power Plants in the United States." *J. Am. Water Resour. Assoc.*; Vol. 43, 2007; pp. 160–169.

[27] Dziegielewski, B.; Bik, T. *Water Use Benchmarks for Thermoelectric Power Generation.* Research Report of the Department of Geography and Environmental Resources. Carbondale, IL: Southern Illinois University, 2006.

[28] Huston, R. *An Overview of Water Requirements for Electric Power Generation.* Water Management by the Electric Power Industry. Gloyna, E.; Woodson, H.; Drew, H., eds. Austin, TX: University of Texas Center for Research in Water Resources, 1975.

[29] Energy Information Administration (EIA). *Form EIA-923 Power Plant Operations Report Instructions, OMB No. 1905-0129*, U.S. Department of Energy, 2011.

[30] U.S. Government Accountability Office (GAO). *Improvements to Federal Water Use Data would Increase Understanding of Trends in Power Plant Water Use.* GAO 10-23. Washington, DC: U.S. Government Accountability Office, 2009.
[31] Solley, W.; Pierce, R.; Perlman, H. *Estimated Use of Water in the United States in 1995.* U.S. Geological Survey Circular 1200. Reston, VA: USGS, 1999.
[32] EIA. *Form 906/920/923: Utility, Non-utility, and Combined Heat And Power Plant Database. 2007 Monthly Time Series.* Washington, DC: U.S. Energy Information Administration, 2010.
[33] NETL. *2007 Coal Plant Database.* Pittsburgh, PA: National Energy Technology Laboratory, 2007.
[34] Leitner, A. *Fuel from the Sky: Solar Power's Potential for Western Energy Supply.* NREL/SR 550-32160. Golden, CO: National Renewable Energy Laboratory, 2002.
[35] Meridian. *Energy System Emissions and Materiel Requirements.* Report to U.S. Department of Energy, Alexandria, VA: Meridian Corporation, 1989.
[36] American Wind Energy Association. *AWEA Estimate Based on Data Obtained in Personal Communication with Brian Roach, Fluidyne Corp.,* 1996. http://archive.awea.org/faq/wwt_environment.html#How%20much%20water%20do%20wind%20turbines%20use%20compared%20with%20conventional%20power%20plants. Accessed November 17, 2010.
[37] Cohen, G.; Kearney, D.; Drive, C.; Mar, D.; Kolb, G. *Final Report on the Operation and Maintenance Improvement Program for Concentrating Solar Plants.* SAND99-1290. Albuquerque, NM: Sandia National Laboratories, 1999.
[38] Kelly, B. *Nexant Parabolic Trough Solar Power Plant Systems Analysis-Task 2: Comparison of Wet and Dry Rankine Cycle Heat Rejection.* NREL/SR-550-40163. Golden, CO: National Renewable Energy Laboratory, 2006.
[39] Sargent & Lundy. *Assessment of Parabolic Trough and Power Tower Solar Technology Cost and Performance Forecasts.* NREL/SR-550-34440. Golden, CO: National Renewable Energy Laboratory, 2003.
[40] Stoddard, L.; Abiecunas, J.; Connell, R.O. *Economic, Energy, and Environmental Benefits of Concentrating Solar Power in California.* NREL/SR-550-39291. Golden, CO: National Renewable Energy Laboratory, 2006.

[41] Viebahn, P.; Kronshage, S.; Trieb, F.; Lechon, Y. *Final Report on Technical Data, Costs, and Life Cycle Inventories of Solar Thermal Power Plants.* 502687. Rome, Italy: New Energy Externalities Developments for Sustainability Project, 2008; pp. 1–95.

[42] WorleyParsons. *Beacon Solar Energy Project Dry Cooling Evaluation.* Report No. FPLS-0-LI-450-0001. North Sydney, Australia: Worley Parsons, 2009.

[43] WorleyParsons. *Analysis of Wet and Dry Condensing 125 MW Parabolic Trough Power Plants.* Report No. NREL-2-ME-REP-0002-R0. North Sydney, Australia: WorleyParsons, 2009.

[44] WorleyParsons. *Material Input for Life Cycle Assessment Task 5 Subtask 2: O&M Schedules.* Report No. NREL-0-LS-019-0005. North Sydney, Australia: WorleyParsons, 2010.

[45] WorleyParsons. *Parabolic Trough Reference Plant for Cost Modeling with the Solar Advisor Model.* North Sydney, Australia: WorleyParsons, 2010.

[46] Kutscher, C.; Buys, A. *Analysis of Wet/Dry Hybrid Cooling for a Parabolic Trough Power Plant.* Golden, CO: National Renewable Energy Laboratory, 2006.

[47] DOE. *Concentrating Solar Power Commercial Application Study: Reducing Water Consumption of Concentrating Solar Power Electricity Generation.* Report to Congress. Washington, DC: U.S. Department of Energy, 2009.

[48] Brightsource_Energy. *Application for Certification, Volumes I and II, for the Ivanpah Solar Electric Generating System.* Submitted to California Energy Commission Docket Unit. Application For Certification (07-AFC-5), August 31, 2007.

[49] California Energy Commission (CEC). *2007 Environmental Performance Report of California's Electrical Generation System.* California Energy Commission Final Staff Report. Sacramento, CA: California Energy Commission, 2008.

[50] Electric Power Research Institute (EPRI). *Water and Sustainability (Volume 2): An Assessment of Water Demand, Supply, and Quality in the U.S.-The Next Half Century.* Technical Report 1006785. Palo Alto, CA: Electric Power Research Institute, 2002.

[51] EPRI; DOE. *Renewable Energy Technology Characterizations.* EPRI Topical Report-109496. Palo Alto, CA: EPRI, 1997.

[52] Mann, M.; Spath, P. *Life Cycle Assessment of a Biomass Gasification Combined-Cycle System.* TP-430-23076. Golden, CO: National Renewable Energy Laboratory, 1997.

[53] Layton, D. "Water-related Impacts of Geothermal Energy Production in California's Imperial Valley." *1979 Annual Meeting of the Geothermal Resources Council.* CONF-190906--18. Livermore, CA: Lawrence Livermore Laboratory, 1979.

[54] Kozubal, E.; Kutscher, C. "Analysis of a Water-cooled Condenser in Series with an Air-cooled Condenser for a Proposed 1-MW Geothermal Power Plant." *Trans. - Geotherm. Resour. Counc.*; Vol. 27, 2003.

[55] EERE. *Geothermal Technologies Program: Geothermal FAQs.* Washington, DC: U.S. Department of Energy Office of Energy Efficiency and Renewable Energy, 2006.

[56] Kutscher, C.; Costenaro, D. *Assessment of Evaporative Cooling Enhancement Methods for Air-Cooled Geothermal Power Plants.* NREL/CP-550-32394. Golden, CO: National Renewable Energy Laboratory, 2002.

[57] NETL. *Estimating Freshwater Needs to Meet Future Thermoelectric Generation Requirements.* DOE/NETL-400/2009/1339. Pittsburgh, PA: National Energy Technology Laboratory, 2009.

[58] Hoffmann, J.; Forbes, S.; Feeley, T. *Estimating Freshwater Needs to Meet 2025 Electricity Generating Capacity Forecasts.* Pittsburgh, PA: National Energy Technology Laboratory, 2004; pp. 1–12.

[59] NETL. *Cost and Performance Baseline for Fossil Energy Plants-Volume 1: Bituminous Coal and Natural Gas to Electricity-Revision 2.* DOE/NETL-2010/1397. Pittsburgh, PA: National Energy Technology Laboratory, 2010.

[60] Feeley, T.J.; Green, L.; Murphy, J.T.; Hoffmann, J.; Carney, B.A. *Department of Energy/Office of Fossil Energy's Power Plant Water Management R&D Program.* Pittsburgh, PA: National Energy Technology Laboratory, 2005; pp. 1–18.

[61] NETL. *Existing Plants, Emissions and Capture – Setting Water-Energy R&D Program Goals.* DOE/NETL-2009/1372. Pittsburgh, PA: National Energy Technology Laboratory, 2009.

[62] Environmental Protection Agency (EPA). *Clean Water Act*, in *Section 316b. Chapter 3*, U.S. Environmental Protection Agency. 2009.

[63] EPRI. *Use of Degraded Water Sources as Cooling Water in Power Plants.* Technical Report 1005359. Palo Alto, CA: Electric Power Research Institute, 2003.

[64] Carter, H.; Schubel, J.; Wilson, R. "Thermally Induced Biological Effects Caused by Once-through Cooling Systems: A Rationale for Evaluation." *Environ. Manage.*; Vol. 3, 1979; pp. 353–368.
[65] Laws, E. *Aquatic Pollution.* 3rd edition. Hoboken, NJ: John Wiley and Sons, Inc, 2000.
[66] Reynolds, J.Z. "Power Plant Cooling Systems: Policy Alternatives." *Sci.*; Vol. 207, 1980; pp. 367–372.
[67] Poumadère, M. "The 2003 Heat Wave in France: Dangerous Climate Change Here and Now." *Risk Anal.: An Off. Publi. Soc. Risk Anal.*; Vol. 25, 2005; pp. 1483–1494.

End Notes

[1] Most geothermal facilities can use geothermal fluids or freshwater for cooling.

INDEX

A

access, 28, 59, 60, 121, 132
accounting, 66, 119, 120
acidic, 47
additives, 13, 55, 56
adjustment, 125
adsorption, 47, 56
age, 12, 51, 54, 67, 94, 119, 144
agencies, viii, x, 2, 3, 5, 10, 25, 28, 30, 31, 32, 50, 51, 53, 57, 60, 62, 63, 64, 91, 93, 94, 96, 111, 112, 117, 135, 136, 141, 144, 145, 153
agriculture, 46, 89
Alaska, 39, 51, 57, 63, 65, 67, 69, 71, 73, 74, 75, 95, 97, 100, 101
algae, 20, 30, 138
American Recovery and Reinvestment Act, 138
American Recovery and Reinvestment Act of 2009, 138
anemia, 13
aquaculture, 59
aquatic life, 114
aquifers, 3, 11, 24, 110, 118
arsenic, 118
assessment, 10, 118, 119
audit, 6, 33, 127, 145
audits, 117, 127
authorities, 9
authority, 8, 9, 24, 25, 26, 47, 57, 63, 79, 115
awareness, 134

B

background information, 53
bacteria, 20, 120, 138
barium, 13, 14, 18
barriers, ix, 107, 108, 111, 123, 131, 132, 134
BED, 36
benchmarking, 127
benefits, 131
benzene, 13
bias, 144
biofuel, 30
biogas, 129
biomass, 146
biopower, 142, 143, 146
blends, 14
blood, 13
blood pressure, 13
boils, 21
bounds, 3, 22
bromine, 76
Bureau of Land Management, 3, 6, 33, 50, 91, 97
businesses, viii, 49, 110
by-products, 133

C

calcium, 13, 21, 47
calibration, 153
cancer, 13
carbon, ix, 15, 46, 84, 138, 140, 143, 148, 149, 152, 154
carbon dioxide, 15, 84, 138
case studies, ix, 108, 111, 135, 144
CEC, 158
Census, 109
challenges, 19, 132
chemical, 20, 27, 28, 52, 53, 54, 55, 56, 58, 137
chemical characteristics, 52, 54
chemical properties, 52, 55
chemical reactions, 20, 58
chemicals, 4, 8, 13, 14, 21, 52, 56, 112, 138
chlorine, 120
cities, ix, 108, 111, 118, 121, 135
City, 47, 96, 98, 101, 103, 122
classes, 8, 46
classification, 10
Clean Water Act, 3, 8, 9, 24, 25, 26, 46, 110, 115, 159
cleaning, 142, 143
climate, 152, 153
climate change, 153
climates, 18, 23
climatic factors, 153
closure, 86
coal, ix, 4, 11, 43, 45, 51, 54, 56, 57, 59, 73, 91, 94, 140, 142, 143, 146, 149, 152
coding, 136
cogeneration, 60
commerce, 46
commercial, 26, 31, 51, 58, 62, 70, 74, 79, 80, 81, 85, 86, 89, 90, 95, 113, 117, 134
communication, 64
communities, 153
community, 112, 134, 137
compilation, viii, 27, 50, 93
complexity, 122, 146
compliance, 55
composition, 27, 28
compounds, 44, 55, 58
conditioning, 113
conductivity, 55, 59
configuration, 143
Congress, 8, 9, 47, 116, 158
conservation, 108, 123, 126, 132, 133
constituents, 21, 55, 56, 58
construction, 27, 31, 47
consumers, 113, 119
consumption, ix, 31, 137, 140, 142, 143, 144, 145, 146, 147, 149, 152, 153, 154
consumption rates, 145
contaminant, 115, 137
contamination, 3, 24, 118
content analysis, 136
cooling, ix, x, 46, 59, 140, 141, 142, 143, 144, 145, 146, 148, 149, 152, 153, 160
cooling process, 146
cooperation, 29
coordination, 132
correlations, 50, 94
corrosion, 13, 15, 20, 56, 58
cost, viii, x, 3, 15, 19, 21, 22, 29, 30, 31, 49, 54, 109, 131, 132, 141, 152
cost saving, 131
covering, 135
crop, 13, 19, 59, 143
crops, 3, 18, 19, 142, 143
crude oil, 51, 67, 68, 69, 74, 75, 76, 77, 78, 79, 80, 81, 82, 85, 86, 87, 88, 89, 90, 91, 94
cultivation, 59
customers, 108, 110, 113, 114, 117, 119, 126, 129, 132, 133, 134, 137

D

data availability, 145
data collection, 66, 153
database, 28, 64, 65, 77, 78, 80, 84, 88, 145
decision-making process, 23
delegates, 9
Department of Defense, 112, 136
Department of Energy, viii, 3, 5, 31, 32, 34, 36, 48, 49, 50, 52, 59, 63, 96, 97, 99,

103, 104, 109, 112, 136, 139, 145, 156, 157, 158, 159
Department of the Interior, 6, 28, 29, 31, 33, 34, 42, 48, 112, 136
Departments of Defense, Energy (DOE), ix, 108
depth, 14, 25, 32, 58, 111, 135
detection, 126
detection system, 126
diesel fuel, 47
digestion, 114, 130, 138
directors, 60
discharges, 3, 9, 17, 23, 25, 47, 57, 62, 63, 70, 73, 74, 75, 89, 95, 115, 120, 130
disclosure, 27
disinfection, 113, 120, 128, 133, 138
dissolved oxygen, 125
distillation, 3, 21
distribution, 108, 113, 121, 129
draft, ix, 2, 31, 108, 134, 144
drawing, 124
drinking water, viii, ix, 5, 8, 16, 20, 29, 33, 46, 47, 49, 57, 107, 108, 109, 110, 111, 112, 115, 116, 117, 118, 119, 120, 122, 123, 124, 126, 127, 128, 129, 131, 132, 133, 134, 135, 136, 137
drought, 141, 153
durability, 125

E

economic development, 112, 136
economic growth, 6
economics, 60
educational materials, 128
effluent, 25, 57, 110, 114, 119, 121, 122, 127, 130, 133
effluent limitation guidelines, 57
electricity, ix, 117, 129, 130, 140, 141, 142, 143, 146, 147, 148, 149, 152, 154
energy conservation, 127
energy consumption, 120, 123, 124, 127
energy efficiency, 108, 123, 124, 125, 126, 127, 128, 132
Energy Policy Act of 2005, 29
energy security, 153
enforcement, 8, 9
engineering, 45, 145
environment, vii, 4, 6, 13, 53, 103, 109, 142, 157
environmental impact, 29, 31, 141, 152
environmental protection, 25, 112, 136
Environmental Protection Agency(EPA), ix, 3, 5, 31, 32, 34, 45, 50, 57, 103, 104, 108, 109, 136, 138, 159
environmental risk, vii, 1
EPA, ix, 2, 3, 5, 6, 8, 9, 13, 15, 16, 17, 24, 25, 26, 28, 29, 32, 33, 46, 47, 57, 63, 75, 87, 103, 104, 108, 109, 110, 112, 115, 116, 119, 125, 126, 127, 128, 132, 133, 134, 136, 137, 138, 159
equipment, 20, 21, 108, 114, 119, 123, 124, 125, 127, 129, 131
equity, 112, 136
evaporation, 18, 23, 51, 57, 58, 70, 77, 82, 89, 95, 146
evidence, 6, 33, 112, 126, 137
expertise, 111, 135
exposure, 13, 58
extraction, 25, 46, 47, 118, 119

F

families, 55
federal agency, 112, 136, 153
federal government, 32, 45, 128, 144
federal law, 8
federal level, vii, 1, 5, 32, 47, 62
federal scientists, viii, 2, 5, 32
fiber, 126
filters, 3, 112, 120, 129
filtration, 20, 21, 30, 31, 120, 122, 137
financial, 132
fish, 28, 46, 116
flooding, 58
flue gas, 142
fluid, 46, 79, 143, 149
food, 113
formation, 4, 7, 8, 12, 14, 16, 20, 30, 33, 51, 52, 54, 55, 56, 58, 95

fouling, 21, 31
fractures, 4, 8, 45
France, 153, 160
freshwater, x, 19, 20, 31, 121, 141, 143, 144, 146, 149, 152, 153, 160
friction, 13, 15, 119
funding, 131
funds, 131
fungi, 20

G

GAO, vii, viii, ix, 1, 2, 17, 33, 45, 107, 108, 110, 112, 113, 115, 137, 157
gasification, 147, 148
geography, 2, 10
geology, 12, 14, 23
GIS, 38, 39, 48
gravity, 118, 121, 137
groundwater, 5, 9, 28, 29, 56, 59, 108, 118, 121, 144
growth, 11, 13, 20, 33, 125
growth rate, 11
Gulf of Mexico, 63, 100

H

habitat, 59
hardness, 21, 47
health, 13, 115
health effects, 115
health risks, 13
heavy oil, 7
homes, 113
House, 1, 4, 107, 109
House of Representatives, 1, 4, 107, 109
human, vii, 1, 4, 13, 137
human health, vii, 1, 4, 13
humidity, 144
hybrid, 143, 152
hydrocarbons, 4, 6, 7, 8, 11, 14, 33, 52, 54, 55, 58, 74, 79, 82
hydroelectric power, 129, 130

I

identification, 40
impact assessment, x, 141
imports, 122
improvements, 8, 129, 131, 153
Indian reservation, 29
Indians, 92
individuals, 111, 136, 139
industries, 46
industry, 5, 28, 29, 30, 31, 32, 75, 111, 112, 134, 136, 144
infrastructure, 108, 109, 113, 119, 123, 125, 131
inhibitor, 56
injections, 78
inspections, 126
institutions, 5, 32
integrity, 25
intervention, 47
investment, 131
investments, 132
ions, 21, 47
iron, 13, 56
irrigation, 3, 15, 18, 19, 20, 23, 24, 26, 27, 43, 47, 59, 91, 121
issues, ix, 3, 6, 28, 29, 33, 45, 107, 112, 136
Italy, 158

J

Jordan, v, 139
jurisdiction, 47

L

lakes, 110
landfills, 114
lead, 59, 119, 144, 153
leaks, 108, 126
legislation, 115
life cycle, 142
lifetime, 51, 53, 54, 55, 67, 94
light, 120, 133, 138

liquids, 8, 12, 81
livestock, 13, 27, 59, 87
local conditions, 153
local government, ix, 86, 108, 112, 136
Louisiana, 5, 17, 27, 33, 50, 65, 66, 71, 80, 94, 102
Luo, 41, 42

M

magnesium, 13, 21, 47
magnitude, 146
majority, 18, 50, 57, 67, 70, 73, 74, 75, 81, 82, 83, 85, 86, 88, 89, 90, 92, 96, 118, 142
man, 8
manganese, 13, 56
manufacturing, 142
mass, 143
materials, 13, 28, 112, 113, 114, 128
matter, 114
measurements, 56
media, 138
median, 145
membranes, 21, 120
meridian, 26, 46
metals, 13
metropolitan areas, 109
Mexico, 5, 12, 14, 17, 18, 26, 28, 30, 33, 39, 48, 65, 69, 71, 75, 83, 84, 99, 101
MFI, 40
microorganisms, 13, 56, 58, 120
migration, 57
mineral resources, 91
Minerals Management Service, 50, 63, 100
Minerals Management Service (MMS), 63
Missouri, 65, 69, 71, 81, 100
MMS, 63, 91, 92
models, 142, 153, 154
modifications, 124
Montana, 12, 14, 36, 40, 44, 65, 69, 71, 73, 82, 99, 100
mortality, 13

N

National Academy of Sciences, 29
National Renewable Energy Laboratory, 155, 156, 157, 158, 159
National Research Council, 18, 19, 34, 45, 47, 48
natural gas, 43, 44, 47, 51, 59, 76, 77, 81, 82, 83, 84, 85, 88, 94, 117, 143, 145, 146, 150, 152
North America, 44
NREL, 139, 140, 155, 156, 157, 158, 159
nutrients, 122

O

oceans, 46, 73
Office of Management and Budget, 137
officials, viii, 2, 5, 9, 10, 15, 16, 17, 19, 21, 24, 25, 26, 31, 32, 33, 34, 46, 47, 112, 119, 121, 122, 128, 136
oil production, 23, 45, 51, 54, 56, 67, 68, 69, 74, 75, 77, 78, 79, 80, 81, 82, 83, 87, 88, 89, 91, 92, 93, 94
oil sands, 4, 45
Oklahoma, 5, 17, 27, 33, 44, 47, 50, 52, 65, 67, 70, 72, 73, 86, 94, 98
OMB, 156
operating costs, 20
operations, vii, 4, 5, 6, 9, 10, 15, 17, 19, 28, 31, 33, 53, 55, 56, 57, 59, 69, 89, 95, 124, 127, 129, 133, 141
opportunities, 23, 127, 133
optimization, 123, 124
organic chemicals, 55
organic compounds, 14, 28, 55
organic matter, 114, 137
osmosis, 21, 22, 31, 128
Outer Continental Shelf, 100
outreach, 53
overlap, 89
oversight, 24
oxygen, 114, 120, 125, 138
ozonation, 120, 122, 133
ozone, 120, 128

P

Pacific, 112, 120, 136
participants, 139
penalties, x, 25, 26, 141, 152
percentile, 145, 147, 148
percolation, 57
permeability, 47
permit, 9, 19, 25, 26, 46, 63, 74, 75, 79, 84, 86, 119, 122
personal communication, 98, 99, 100, 101, 102, 103
Petroleum, 30, 38, 39, 40, 42, 43, 44, 48, 54, 97, 98, 101
pH, 21, 47
physical characteristics, 53
pipeline, 22, 108, 126
plant growth, 59
plants, 59, 108, 115, 117, 119, 120, 121, 122, 127, 128, 137, 138, 141, 143, 145, 146, 152, 153
police, 131
policy, ix, x, 137, 140, 141, 154
pollutants, 9, 25, 46, 114, 115
pollution, 9, 116
ponds, 3, 18, 26, 27, 51, 58, 62, 70, 77, 82, 89, 95
population, 109, 116, 125
population growth, 125
portfolio, 6, 142
potential benefits, 124
power generation, ix, 59, 140, 142
power plants, x, 141, 144, 145, 153
precipitation, 116
producers, vii, viii, 2, 3, 4, 5, 9, 11, 14, 15, 16, 17, 18, 19, 20, 21, 22, 23, 24, 25, 26, 27, 28, 29, 33, 46, 47, 49, 50, 58, 66, 67, 77, 94, 95
profit, 54
project, 28, 30, 31, 37, 38, 39, 40, 41, 48, 59, 60, 86, 97, 121
propagation, 116
protection, 116, 131
prototype, 30
public awareness, 123, 131
public education, 126
public health, 8
pumps, 113, 120, 124, 125, 131

Q

questionnaire, 60, 62, 63

R

radioactive waste, 46
reality, 19
recommendations, ix, 2, 108
recovery, 8, 12, 14, 16, 17, 23, 46, 51, 55, 57, 58, 59, 62, 66, 70, 74, 75, 76, 77, 78, 79, 80, 81, 82, 83, 84, 85, 86, 87, 88, 89, 90, 92, 95, 129
recreation, 46, 146
recreational, 116
recycling, 86
redistribution, 9
regulations, 3, 5, 9, 19, 22, 25, 26, 27, 29, 115, 122, 133, 153
regulatory agencies, 26
regulatory changes, 122
regulatory requirements, 57, 119, 133
reliability, ix, 19, 32, 140
remediation, 30
renewable energy, 123, 128, 129, 130, 131, 142
renewable energy technologies, 142
requirements, x, 9, 18, 20, 23, 27, 31, 66, 115, 116, 122, 125, 133, 141, 142, 143, 146, 154
researchers, 10, 112, 136
reserves, 4, 6, 45
resistance, 46, 132
resolution, x, 141
resource management, 9
resources, ix, x, 28, 29, 33, 45, 47, 51, 63, 64, 67, 69, 94, 96, 109, 140, 141, 152
response, 25
revenue, 92, 132
reverse osmosis, 21, 22, 26, 31
risk, vii, 1, 13, 23, 58, 115

risk management, 23
risks, 6, 13, 153
royalty, 45
rules, 47, 57
rural areas, 152

S

safety, 133, 138
saline water, 144, 153
salinity, 55, 56, 57, 59
salt concentration, 18, 19, 21, 47
salts, vii, 4, 13, 19, 21, 22, 47
saltwater, 54
savings, 124, 126, 129, 131, 132
scaling, 21, 56
scarcity, 109, 145
Science Advisory Board, 29
scope, 60, 62, 154
security, 138
sediment, 43
sedimentation, 112, 114
seeding, 9
semi-permeable membrane, 21
sensors, 126
services, viii, 107, 108, 109, 111, 116, 119, 120, 128, 131, 133, 137
sewage, 114, 116
shellfish, 46, 116
sludge, 114, 120
sodium, 13, 21, 47, 59
soil type, 19
solution, 84, 123, 127
solvents, 30, 56
South Dakota, 65, 67, 69, 72, 73, 87, 94, 101
specialists, ix, 107, 108, 111, 116, 117, 118, 119, 120, 122, 123, 124, 125, 126, 127, 128, 129, 131, 132, 133, 134, 135, 136, 137
species, 59
staff members, 96
staffing, 123
stakeholders, 5, 11, 12, 13, 14, 22, 28
state authorities, 3, 24
state regulators, 17, 19, 22
state regulatory officials, viii, 2, 5, 32
states, vii, 1, 2, 3, 5, 6, 9, 10, 11, 16, 17, 18, 19, 24, 25, 26, 27, 29, 32, 33, 50, 51, 52, 60, 62, 63, 64, 66, 67, 68, 70, 73, 78, 80, 87, 93, 94, 95, 96, 97, 112, 132, 136
statistics, 66, 90, 91, 117, 145
stimulation, 143
stock, 82
storage, 27, 47, 57, 79, 84, 143, 148
stormwater, 116, 128, 137
stratification, 79
strontium, 13
structure, 47, 59
style, viii, 107, 109
sulfur, 47
supplier, 127
suppression, 82
surface area, 58
susceptibility, 59
Switzerland, 154

T

tanks, 113
team members, 97
technical comments, ix, 2, 31, 108, 134
techniques, 5, 15, 31, 33, 112, 132, 136, 143
technologies, viii, ix, x, 3, 7, 19, 21, 22, 27, 30, 107, 108, 111, 115, 118, 120, 122, 123, 126, 127, 128, 131, 132, 134, 138, 140, 141, 142, 143, 144, 145, 146, 149, 152, 153, 154
technology, 21, 29, 31, 125, 126, 128, 133, 142, 143, 144, 145, 146, 149, 153
temperature, 31, 58, 141, 143, 153
territory, 122
testing, 3, 24, 25, 30
time frame, 11
toluene, 13
total energy, 6, 116, 120, 127, 133
total product, 76, 91
toxicity, 56, 59
tracks, 76
transmission, 130, 154

transport, 46, 116
transportation, 23
treatment methods, 3, 15, 20, 22, 30, 32, 109
tribal lands, 50, 53, 64, 67, 91, 92, 93
TVA, 156

U

U.S. Army Corps of Engineers, 112, 136
U.S. Department of Agriculture, 91
U.S. Department of the Interior, 63, 91, 100
U.S. Geological Survey, 3, 6, 33, 43, 48, 100, 109, 112, 136, 139, 154, 156, 157
universities, 29
urban, viii, 107, 108, 109, 110, 111, 112, 116, 117, 118, 119, 120, 122, 123, 124, 126, 129, 131, 133, 134, 135, 137
urban areas, 109, 116, 129
urban life, viii, 107, 109
urban population, 111
USA, 44
USGS, 3, 12, 28, 30, 44, 47, 109, 141, 142, 145, 153, 154, 156, 157

V

vapor, 56, 143
variations, 124, 144, 153
vegetation, 9
viscosity, 7

W

Washington, viii, ix, 45, 47, 49, 97, 98, 99, 100, 101, 102, 103, 108, 111, 118, 122, 126, 135, 137, 138, 156, 157, 158, 159
waste, vii, viii, 4, 8, 25, 46, 49, 53, 58, 84, 137, 144
waste disposal, 137
wastewater, viii, ix, 8, 47, 107, 108, 109, 110, 111, 112, 114, 115, 116, 117, 119, 120, 121, 122, 123, 124, 125, 127, 128, 129, 131, 132, 133, 134, 135, 136, 137, 138, 144, 152
wastewater utilities, ix, 108, 123, 127, 129, 133
water chemistry, 28
water policy, 141, 152
water quality, 5, 9, 10, 14, 18, 23, 26, 28, 30, 32, 56, 114, 116, 119, 122, 133, 138
water quality standards, 9, 18, 26, 119, 122, 133
water resources, ix, 29, 47, 52, 140, 141, 142, 152, 153, 154
water supplies, 8, 23, 52, 115, 116, 121
watershed, 118, 153
websites, 63
wetlands, 29
White Paper, 35, 48, 53, 104
wildlife, 13, 46, 59, 116
wind speed, 144
wind speeds, 144
Wisconsin, 112, 136
withdrawal, ix, 140, 142, 143, 144, 145, 146, 148, 150, 153, 154

Y

yield, 15, 59